Revolutionary Networks

Studies in Early American Economy and Society from the Library Company of Philadelphia
Cathy Matson, Series Editor

Revolutionary Networks

The Business and Politics of Printing the News, 1763–1789

JOSEPH M. ADELMAN

Johns Hopkins University Press
Baltimore

Printed in the United States of America on acid-free paper
2 4 6 8 9 7 5 3 1

Johns Hopkins University Press
2715 North Charles Street
Baltimore, Maryland 21218-4363
www.press.jhu.edu

Library of Congress Cataloging-in-Publication Data

Names: Adelman, Joseph M., 1980– author.
Title: Revolutionary networks : the business and politics of printing the news, 1763–1789 / Joseph M. Adelman.
Description: Baltimore : Johns Hopkins University Press, 2019. | Series: Studies in early American economy and society from the Library Company of Philadelphia | Includes bibliographical references and index.
Identifiers: LCCN 2018032687 | ISBN 9781421428604 (hardcover : alk. paper) | ISBN 9781421428611 (electronic) | ISBN 1421428601 (hardcover : alk. paper) | ISBN 142142861X (electronic)
Subjects: LCSH: Printing industry—United States—History—18th century. | United States—History—Revolution, 1775–1783—Press coverage.
Classification: LCC Z244.6.U5 A34 2019 | DDC 686.20973/09033—dc 3
LC record available at https://lccn.loc.gov/2018032687

A catalog record for this book is available from the British Library.

Special discounts are available for bulk purchases of this book. For more information, please contact Special Sales at 410-516-6936 or specialsales@press.jhu.edu.

Johns Hopkins University Press uses environmentally friendly book materials, including recycled text paper that is composed of at least 30 percent post-consumer waste, whenever possible.

For Sarah

CONTENTS

SERIES EDITOR'S FOREWORD

In this addition to the series Studies in Early American Economy and Society, a collaborative effort between Johns Hopkins University Press and the Library Company of Philadelphia's Program in Early American Economy and Society (PEAES), Joseph M. Adelman examines the role of printers in the Revolutionary generation. His meticulous research uncovers who these vital revolutionary figures were and how they had a profound impact on the intellectual and action platforms of Patriots and Loyalists; Adelman also looks at how the process of printing, including its trade and commercial practices, contributed directly to political debates. Most printers, Adelman argues, were neither staunch Patriots nor shrill Loyalists; rather, they were measured in their responses to the imperial crisis or silent altogether about the unfolding rebellion. But together, whether shrill, determined, measured, or silent, these printers held responsibility for constructing and distributing news during an extraordinarily newsworthy era.

While some printers wrote in a politically rhetorical vein during the Revolutionary era, many others focused on their business role and were more removed from directly formulating political views. These latter edited news items received in their printing establishments, which then appeared as publications, in newspapers, pamphlets, broadsides, sermons, and other printed formats. Printers also played a central role in distributing print materials to both local and distant communities. Indeed, because the printing trade relied on long-distance information as its key commodity, printers built networks with other printers and distributors not only along the Atlantic colonial corridor but in the Caribbean and across the Atlantic Ocean as well. Some of these networks were quite durable, especially those that originated out of family, partnership, or master-apprentice relations. Perhaps this will be unsurprising to readers familiar with the significance of family trust and investment networks among merchants and small entrepreneurs throughout the Atlantic World. Equally unsurprising, many networks of printers remained fragile business relationships, and some failed to flourish. What Adelman reveals for the first time, however, is that printers could not thrive as independent businessmen outside of deep connections not only to writers, distributors, producers of paper and ink, and consumers but also to similarly situated printers in distant places who expanded the orbit of news collection and the readerships of North American printers.

As Adelman insists, the Revolutionary War scattered printers, technologies of print, and readerships. Many printers fled their homes and local shops, with the result of sometimes spreading the impact of printed news into the interior and sometimes destroying a printer network altogether. Those who set up print businesses in the interior during the Revolution struggled mightily to obtain paper and ink and to reestablish distribution webs. But for those who prevailed against the Revolution's difficulties, the end of the war introduced new opportunities for existing printers as well as immigrant printers with a few resources.

Indeed, by the mid-1780s the printing trade began a rapid expansion that would continue for decades. More than ever before, printers contributed pragmatically to requirements for commercial and legal forms, fulfilled custom orders for broadsides, and put out almanacs, all of which would generate crucial revenue. While some published pamphlets, only a few undertook the resource-eating projects of book production; instead, until at least the end of the 1780s, most printers imported the books they wished to make available to readers (or provided them to discerning readers on special order). Moreover, the political culture of printers' enterprises foregrounded articulate and confident men who hitched their futures to advocating political positions and policies they believed fulfilled the promise of the Revolution. And these positions and policies were best explicated in the columns of newspapers, which expanded slowly from their modest beginnings as four-page compilations of news published weekly to longer and more frequent publications during the post-Revolutionary generation involving printers in the roles of editor and publisher, men with substantial control over the political content of his newspaper. Readers who venture into Adelman's innovative and thought-provoking portrayal of Revolutionary-era printers will be handsomely rewarded with numerous insights into how North Americans knew what they knew.

CATHY MATSON
Richards Professor of American History, University of Delaware, and
Director, Program in Early American Economy and Society,
Library Company of Philadelphia

If you're like me, when you picked up the book, whether in a bookstore, a library, or even (hopefully!) when you opened the box or bag after buying it, you turned first to this page to see whom I've thanked. If so, welcome to my book. In my experience, people turn to the acknowledgments first out of intellectual interest (With whom did he work? What archives did he visit?). Some interest can better be categorized as prurient curiosity (Did he say something super awkward about his spouse?). And a few of you are just checking to see if your name is here. If you think it should be and it's not, let me offer my apologies in advance, and please know that I am grateful for your support.

The research and writing that this book represents spans nearly two decades, during which time I've been very fortunate in finding mentors who have devoted significant time and energy to my intellectual development and this project in particular. Cathy Matson, director of the Program in Early American Economy and Society, has never flagged in encouraging and shepherding me to complete the manuscript, revise it, and work through the publication process. The book developed from a dissertation I completed at Johns Hopkins University, where I had not one but two generous and watchful advisers, Toby Ditz and Phil Morgan. Their styles were very different, and their advice occasionally conflicted, but the work is far better for having both of them. On top of reading draft after draft of chapters, pressing me to push my conclusions further, and editing my writing (thoroughly), they supported me through a difficult job market and encouraged me to pursue my interests. Others at Hopkins, including John Marshall, Michael Johnson, Mary Ryan, and Megan Zeller, also provided generous support in various ways, both on this project specifically and in helping me navigate my way through graduate school and the job market. As an undergraduate, I wrote my senior thesis on American national identity during the Revolution with Joyce Chaplin, who has remained a mentor (and helped me get tickets to see *Hamilton*). Her clear advice and uncompromising standards pushed the twenty-one-year-old me to work harder, and now we can enjoy a shared love of Benjamin Franklin. And I can trace the intellectual origins of this book all the way back to a seminar on "Communication in the New Nation" that I took my sophomore year of college with Catherine Corman, in which I wrote a research essay on the Boston Committee of Correspondence and its tactics to create an intercolonial political network. (Really.)

Many other scholars have read and contributed to the project, and it is much richer for their insights and suggestions. Richard John and Jeff Pasley in particular have devoted significant time to conversation and advice regarding the post office, early American journalism and communications, and the job market, among other things. I am very grateful to Andy Shankman, Dael Norwood, and Michelle Craig McDonald, who read the entire manuscript and conducted a workshop with me during my time in Philadelphia. They offered insightful comments that helped me develop my thinking in a number of key areas. I would also like to thank Mark Schmeller, who read the entire manuscript for the press and wrote a very gracious review with numerous helpful suggestions. I appreciate the comments and perspectives from those who have reviewed my work, including Catherine O'Donnell, David Waldstreicher, David Henkin, Rob Desrochers, Konstantin Dierks, Richard D. Brown, and Robert Gross. Dan Richter, director of the McNeil Center for Early Americans Studies, has offered intellectual and material support for many years and helped make Philadelphia an intellectual home away from home. He also provides the bar by which I judge my academic puns. The editors at Johns Hopkins University Press, including Laura Davulis, Lauren Straley, and Juliana McCarthy, have been helpful and encouraging throughout the publication process. I also would like to thank Brian MacDonald, who copyedited the manuscript.

Funding for humanities projects is under threat in the current political environment, and I am deeply grateful to the institutions that have supported my work over the years. If you like this book—or even if you don't but think books like it should exist—please support the humanities, both private entities and our many government-supported institutions. First and foremost, the Program in Early American Economy and Society (PEAES) at the Library Company of Philadelphia, as well as the Library's general fellowship funds, offered me intellectual space and time to work. I also received long-term funding from the National Endowment for the Humanities to complete research and writing at the American Antiquarian Society. In addition, a number of other institutions have supported the research at various stages: the David Library of the American Revolution, the American Philosophical Society, the American Antiquarian Society, the Gilder-Lehrman Institute of American History, the Doris B. Quinn Foundation, Johns Hopkins University, and Framingham State University. And if we're being honest, I also need to acknowledge crucial assistance from chocolate chip cookies, Reese's peanut butter cups, peanut M&M's, coffee, and several stashes of holiday candy that my kids forgot existed. (Coming soon to my life: more salad!)

The archives and cultural institutions above offered not only funding but also the expertise of dedicated librarians and archivists. Jim Green, librarian at the Library Company of Philadelphia, is a sage when it comes to the early American printing trade—there are many errors that do not appear in these pages thanks to his careful reading and consistent advice. He has also become a friend capable of steering me away from the task at hand to nerd out for an afternoon about Benjamin Franklin. I've also greatly benefited at the Library Company from the help of Connie King, Linda August, Edith Mulhern, and Rachel D'Agostino. Like my children, I love my archives equally, and the American Antiquarian Society has become my intellectual home base in Massachusetts. I want to thank everyone who has worked there since I started visiting in 2005, and in particular Ashley Cataldo, Paul Erickson, Ellen Dunlap, Caroline Sloat, Joanne Chaison, Elizabeth Watts Pope, Kim Toney, and Molly O'Hagan Hardy. Many others have helped along the way, including Meg McSweeney and Kathie Ludwig of the David Library of the American Revolution; John Pollack at the Van Pelt Library at the University of Pennsylvania; John Overholt at Houghton Library; and Earle Spamer at the American Philosophical Society. At the Whittemore Library at Framingham State, Danielle Lamontagne and Kieran Shakeshaft worked tirelessly and with great success to complete my continual interlibrary loan requests. Finally, for the past several years I have been privileged to be part of the team at the Omohundro Institute. I am deeply grateful to Director Karin Wulf for affording me the opportunity to help shape the digital profile of the OI and for her support of my intellectual endeavors. Liz Covart trusted me enough to guest-host an episode of *Ben Franklin's World*, and I deeply value her advice and friendship, even if we disagree about New England baseball. I'd also like to thank Martha Howard, Holly White, and the entire OI team for their support.

People outside of Johns Hopkins often notice how tight-knit the graduate cohorts of early Americanists become, to the point of coining some less-than-savory nicknames about us. The rumors are true even if the nicknames are a little cutting. Among my cohort, Katie Jorgensen Gray deserves first mention. She literally fed me within hours of my arrival in town in August 2004, and then kicked my butt for the next six years to turn me into a better historian (with admittedly mixed results). A special thanks goes also to Katie's parents, Anne and Craig Jorgensen, who welcomed me into their home in the Philadelphia area on research trips (including a memorable few weeks that overlapped with an early season of *Dancing with the Stars*) and have looked out for me and our family ever since. Eddie Kolla and James Roberts likewise became fast friends, trusted confidants, and wry observers of American society from a Canadian perspective. Matt Bender has been a mentor, friend, guide to Baltimore, and a ready companion for a baseball game wherever we find ourselves. Sara Damiano, my academic "little sibling," has carefully read a number of chapters and offered advice on a variety of subjects—including, with Katie, encouragement to think more carefully about the role of women and gender in the printing trade. Jessica Choppin Roney graciously read a chapter and soothed my anxious mind as I was in the final editing process. The Atlantic history seminar is a mainstay of Monday evenings at Hopkins. It greatly encouraged my intellectual growth but more importantly created a space where I found my closest friends at Hopkins and beyond. I'll surely miss someone, but among Hopkins folks I would like to thank particularly Zara Anishanslin, Claire Cage, Kelly Duke-Bryant, Rob Gamble, Stephanie Gamble, Claire Gherini, Jonathan Gienapp, Amanda Herbert, Katherine Hijar, Cole Jones, Carl Keyes, Gabriel Klehr, Caleb McDaniel, Kate Murphy, Greg O'Malley, Justin Roberts, the late Leonard Sadosky, David Schley, Jessica Stern, Molly Warsh, and Jessica Valdez. I've also been very fortunate in the many scholars with whom I've been able to talk about history at conferences, research fellowships, not to mention on social media. My thanks to Sari Altschuler, Rick Bell, Lori Bihler, Lissa Bollettino, Christine DeLucia, John Demos, Carolyn Eastman, Andrew Fagal, Hannah Farber, Becky Goetz, Adam Gordon, Rachel Herrmann, the late Jack Larkin, Sharon Murphy, Ken Owen, Chris Parsons, Josh Piker, Yvette Piggush, Gautham Rao, Brett Rushforth, Virginia Rutter, and last but not least, the team at the *Junto*, the blog where my prose evolved considerably from its graduate school formalism.

The above I met in a professional context. From here on we get progressively more personal. Dannah Rubinstein and Teresa Fazio each deserve some sort of award because they've stuck with me since high school. We've been through a lot in life, and I'm all the better for their friendship. Michèle Cella was my Latin teacher in high school and taught me more about how the English language works than anyone else. After so many years, I am very glad that she is both magistra et amica. Joe and Laura Blasi, Stacy and Paul Fonstad, Maggie Gardner, Ben Miller, Sara Morrison, Clair Waterbury, and Allison and Bernie Yutesler have all provided support, grounding, and much-needed distraction from work. For the past several years I've been fortunate to be part of the Jameson Singers; the camaraderie of old friends and new and the joy of joining together in harmony have sustained and nourished me.

My family has been steadfast throughout my career, including my parents, Alice and Richard Adelman, my brothers Michael, Brian, and Patrick, and my sisters-in-law Melinda and Shayna. Becoming a history professor has occasionally made me the object of humor, but that's how my family shows they care. (They must care a *lot.*) We've now also been joined recently by my nephew Benjamin and niece Annabel, who are joyful diversions. My grandmother, Claire Adelman, thinks I can do no wrong (so long as I call with some regularity and bring the kids to visit), and tells me so in ways that are still meaningful. My other grandparents—Murray Adelman and Joseph and Edythe Jazzo—have long since passed away, but I know they would be proud to hold this volume. My in-laws, Dan and Kathleen Mulhall, have been indefatigable supporters and welcomed me into their family with open arms, not to mention their large extended families. Kathleen in particular has devoted an incredible amount of time to supporting Sarah and me in myriad ways. We are eternally grateful for the help. My children, Nicholas, Elizabeth, and Rebecca, joined us at various stages of the project. Both individually and as a group they keep me on my toes, make me work harder, and distract me in all the ways that kids can. The good news and bad news is they'll keep on growing up even as this project comes to a close. Then there's Sarah. She has her own career, in which she is a marvel of productivity and efficiency. And without her this book would simply not exist. She has by turns offered editing, coaching, support, and the occasional reminder to stay on task. She also took on, especially in the last months of editing, much of my share of family responsibilities. I cannot offer back the time for which I asked, cajoled, or begged (though I will try in the months ahead), so instead I offer the product of that time.

Revolutionary Networks

Introduction

As Isaiah Thomas approached his sixtieth birthday in January 1809, he reflected on his career and role in the previous half-century of American history. Apprenticed as a young orphan into the printing trade, he had worked all over Britain's North American colonies as a teenager, finally settling back in his hometown of Boston as an adult. There he played a central role in fomenting resistance against British imperial policies as a young printer. Affiliating with the Sons of Liberty and the Boston Committee of Correspondence—two of the most radical groups in all of the colonies—Thomas published the *Massachusetts Spy* with a strong anti-imperial voice. And on April 18, 1775, just before the British marched on Lexington and Concord, Thomas fled the city to escape arrest by authorities.[1]

With his press, type, and little else in tow, Thomas traveled forty miles west to Worcester, a small farming town in central Massachusetts. There he reestablished his press, resumed publication of the *Spy*, and embarked on a career that would take him from newspaperman to one of the leading book publishers in the early United States. Flush with the funds from his many bookselling and publishing partnerships, Thomas became a philanthropist, collector, and antiquarian, which eventually led him in 1812 to found the American Antiquarian Society. As part of the work that led to the creation of the AAS, Thomas collected as many publications from the Revolutionary era as he could from his compatriots in the cause in order to write a history of printing in America, from its earliest days to the Revolution. In addition, he corresponded with many of his former colleagues about their experiences and remembrances. Several of them, in particular William Goddard and William McCulloch, sent lengthy sets of suggestions for revision for an anticipated but never realized second edition.[2]

Published in 1810, *The History of Printing in America* represented the culmination of nearly a decade of research on almost two centuries of printing. In the book, Thomas portrayed the printing trade as a noble one, tracing its roots from North America back to Europe and Johannes Gutenberg. Its impact, he claimed, was all-encompassing: "Whatever obscurity may rest upon the origin of Printing," he wrote, "the invention has happily been the mean of effectually perpetuating the discovery of all other arts, and of disseminating the principles by which they are accomplished."[3] The book got more grandiose from there.

Thomas used his *History* to extol the virtues of printing and the importance of printers as a group in America from the arrival of the first printing press in Massachusetts in the 1630s "down to the most important event in the annals of our country—the Revolution."[4] He profiled as many printers as he could, starting with Stephen Daye, who manned the first press in Cambridge. He knew many of the more recent participants in the trade, so their profiles took on a personal and sometimes political tone. For printers such as Benjamin Edes and John Gill, Thomas praised their contributions to the anti-imperial effort. As publishers of the *Boston Gazette*, Edes and Gill were central to the anti-imperial effort, and Thomas described them in glowing terms: "Edes was a warm and a firm patriot, and Gill was an honest whig."[5] As for those who were Loyalist or insufficiently patriotic, Thomas could be merciless. He noted, for instance, that "even the royalists censured" James Rivington, the most famous Loyalist printer in North America, "for his disregard for the truth."[6] After the war, Rivington stayed in New York and continued to publish his *Gazetteer*—as far as Thomas was concerned, the paper was known to be "a wolf in sheep's clothing."[7]

The *History of Printing in America* set the bar for discussing printers' roles during the American Revolution—and set it quite high. Thomas took David Ramsay's 1789 observation that "in establishing American independence, the pen and the press had merit equal to the sword" and drew it to its logical end.[8] For Thomas, the "freedom of the press was the first, and one of the greatest agents in producing our national independence."[9] Printers were heroes, in other words, and their publications integral. Historians have had a hard time shaking that portrayal of the press—this book joins well over 150 books in using the Ramsay quotation, for example—and for good reason. Newspapers are perhaps the most plentiful and easily accessible resource from the Revolutionary era, and the image of the ink-stained printer laboring over the literature of the Revolution is romantic. Yet it obscures as much as it reveals.

The present study offers a central role to printers and their publications. Yet it does so not as a measure of guild solidarity, as was the case with Thomas, or simply by applauding the printed texts of resistance, as Ramsay did. Rather, the book interrogates closely exactly how the process of printing—its trade and commercial practices—shaped the content of political debates. Further, it offers a more critical eye than early chroniclers such as Thomas and Ramsay. *Some* of the printers in this study were heroic, at least from the perspective of the Patriot cause. Others, however, were staunch Loyalists, holding firm with so many of their neighbors to the idea that, they owed allegiance to the British Crown, and that therefore to transgress that commitment was treason. Most fell somewhere in the middle. At times they supported anti-imperial arguments but perhaps saw rebellion as a step too far. And then there's a final group: the silent, those who either ducked from political debate altogether or simply were not important enough as printers to leave more than a trace in the historical record. This volume encompasses all of these groups in order to see them as part of the same news ecosystem.

I argue that printers played a crucial role in the formation and shaping of political rhetoric during the American Revolution. As part of the ordinary course of business, printers learned how to edit and select items for publication—most importantly paragraphs for newspapers but also texts for publication as pamphlets, broadsides, and other printed material—and how to distribute that information to distant locations. Because the printing trade relied on long-distance information as its key commodity, printers developed connections with their comrades in other towns in colonies along the Atlantic seacoast and the Caribbean as well as across the Atlantic. Local news could easily be transmitted by word of mouth in the small communities of British colonial North America, so printers worked to publish news that had the greatest impact—that from London in particular and later, during the 1760s and 1770s, from other colonies. These connections were sometimes substantial and long-lasting, especially when they developed out of a master-apprentice, partner, or kin relationship. In other cases the relationship was ephemeral or formal, distant and business-like but with no other connection.

During the imperial crisis, the connections that printers had begun to develop as part of their trade proved not just indispensable but in fact necessary and vital to the Patriot cause. For Loyalist printers, who made far more connections to government officials than to other printers, these connections proved difficult to leverage into the ability to circulate news. The Revolutionary War, which dragged on for eight years and scattered the inhabitants of nearly every Atlantic port town (and therefore nearly every center of printing) proved exceedingly difficult for printers. For many, the war forced them to flee their homes and businesses, either to relocate further inland (sometimes temporarily, sometimes permanently) or to close their businesses altogether. Those who remained in business further had difficulty gathering materials and news because of the wartime interruptions to communication and commerce.

At the conclusion of the war, the printing trade began a rapid expansion that would continue for decades as Euro-Americans began to push their settlements into the North American interior. Emboldened by their self-proclaimed role in effecting American independence, many printers took a lead role in promoting national unity. Again because of their interest in long-distance connections, printers as a group became among the strongest supporters of the Constitution of 1787, which promised to knit the thirteen states together more closely than had the Articles of Confederation. By the end of the 1780s, therefore, printers could claim that they had created the conditions for the Revolution, promoted its success, and helped to stitch together the United States at the war's conclusion. These factors contributed to the key role that printers and newspaper editors would play in the political system of the early Republic.[10]

Printers produced a range of material from their offices. All did what was called job printing, which entailed essentially custom orders for broadsides and other small items, as well as printing blank forms for commerce and the law. Nearly everyone published an almanac, which was the most popular print medium in the colonies and a sure steady seller. And many published pamphlets, longer works that could be produced quickly and were typically sold unbound. Printers printed few books because they consumed so many resources; instead, booksellers imported nearly all books from abroad until after the Revolutionary War. Much of the output of these colonial offices focused on religion and depended on ministers, who liked to publish sermons they thought of enduring public importance. If a printer was lucky, rival ministers would use his press to engage in debate, keeping his office busy and boosting sales. Because demand for these services ebbed and flowed, printers relied on newspapers as a steady product to bring in revenue and to make the public aware of printers' skills, products, and availability for other jobs.[11] During the first century and a half of printing in America (1639 to 1790), newspapers accounted for nearly 80 percent of all imprints—even though the first multi-issue newspaper appeared only in 1704.[12] In the years before 1775, these newspapers were largely concentrated in the three largest towns in

North America: Boston, New York, and Philadelphia. Before the Revolution, all but a handful published on a weekly basis, typically about four pages on a single sheet. The typical newspaper comprised news items, advertisements and notices, government proclamations, announcements, and essays. The news items included correspondence and other reports not only from the town in which the newspaper was printed but also from Europe and elsewhere in America. The printer was in almost all cases also the newspaper's editor and publisher. He decided what news was published and, in concert with his journeymen and apprentices, how to lay out the newspaper. Acting within the confines of acceptable debate, the printer therefore exerted enormous control over the political content of his newspaper.

No process took up as much of a printer's time as the weekly task of compiling, editing, and composing the newspaper. It was central to a successful printing business because it was often the main source of an office's revenue and also the means of advertising both the printer's other goods and his skills at the press. Printers compiled information from various sources, including items in the pages of a newspaper, government proclamations, journals of legislative activity, and advertisements for goods and services. They had to sift through an enormous amount of information, clipping stories from other newspapers, organizing oral reports into newsworthy paragraphs, and excerpting letters from friends and associates. To obtain their information, printers had to maintain contacts with government officials, prominent merchants and ship captains, and other well-placed political sources not only within their localities but from around the colonies and across the Atlantic Ocean. Printers, then, stood at the convergence point for both political news and commercial information that had their fullest expression in the newspapers they printed.

Nearly all newspapers in the late colonial era were published weekly because the work involved in producing even a four-page edition could take up the better part of a week. The newspaper's constituent parts, however, required different degrees of attention each week. Of the three basic categories of text—the masthead and colophon, advertising, and news (very loosely defined)—the first was evergreen, that is, it appeared every single week in the same form for months or years at a time. For this text—including the title of the newspaper, its motto, and perhaps a small image for the top of the first page, and some information about the printer and subscriptions for the bottom of the last page—printers could set the type once, tie it together with string, and reuse it each week without having to redo the composition of the text. A similar process occurred for advertising, which both advertisers and printers preferred to have run for several weeks in a row. If an advertiser paid up front for a several-week run, the printer could similarly save labor by composing the ad with his type, binding it with string, and setting it aside for reuse the following week. For the news paragraphs, however, printers and their employees needed to set new type each and every week. To do that, a printer had to use a composing stick in which he set an individual letter, punctuation mark, or blank space in a line and did so upside down and backward so that the line would appear correctly on the page when imprinted as a mirror image.

The technological constraints inherent in the printing process embedded in a fixed order the presentation of imperial, regional, and local relationships through the material layout of the newspaper. A standard four-page newspaper would be printed on a single sheet of paper. Size varied, but most sheets were in the range of a modern paper size of 22 by 17 inches, which when folded once, made pages of about 11 by 17.[13] The sheet would be printed in two runs through the press, first for the outside pages (i.e., 1 and 4) and then for the interior pages. That process took about two days: to prevent ink from bleeding, the paper was dampened before being placed on the press, and the ink required several hours to dry before the paper could be redampened for the second run.[14]

That technological reality—print pages 1 and 4 together and then 2 and 3 together, plus waiting a day between runs—shaped how printers laid out text on the page. To begin filling in text, they therefore tended to work diagonally from the top left of page 1 and the bottom right of page 4. They first printed news on hand early in the week, which meant that page one often featured official proclamations, essays, and imperial news recently arrived by ship. Inside the newspaper, the news was laid out in increasing order of proximity to the town in which the printer worked, which typically left a small amount of space for local news. The remainder of the newspaper—sometimes as much as two entire pages—was devoted to advertisements. These were absolutely crucial to the newspaper and often to a printer's overall business because they usually provided the most reliable source of revenue.[15] Almost all newspapers also organized their news geographically. That is, the first news to appear would be from Europe (almost all of it filtered through London, regardless of its origins). Then the printer would lay out news in descending order of distance from the site of his office. This news was reproduced from other newspapers that the printer received in his office, either through his role as a postmaster or via subscription. The printer did not produce this text, but he did exercise editorial judgment to select which of the many paragraphs in a given newspaper he wanted to reprint in his own. The last portion of the news to be laid out, usually appearing on the third page of the newspapers, was the "local" news from the town. Often less than a single column, this was the only text that was produced for the issue of a newspaper. As the editors and compilers of all of this information, printers played in active role in filtering and managing the process of news production.

The May 26, 1763, issue of the *Pennsylvania Journal; and Weekly Advertiser* offers a good example of a typical newspaper (figs. 1 and 2). Lacking any long essays for that issue, printer William Bradford instead loaded the front page of the issue with advertising from a variety of sources: a bookseller in Wilmington, Delaware; a man seeking a runaway servant; several sellers of distilled spirits; lists of imported goods for Atlantic merchants; and an announcement for the auction of a parcel of land. The advertisements continued on the entire fourth page and also took up more than two-thirds of page 3. As a group, they represented a broad spectrum of the economic life of Philadelphia and the surrounding region, including not only the Pennsylvania hinterlands but also southwestern New Jersey and northern Delaware (not unlike today's metropolitan newspapers).

On the interior pages, Bradford first printed a plan for "providing and employing a considerable number of those seamen and non-commission officers, who shortly will, by the war, be discharged out of the Navy service." The proposal, signed merely "C.D.," argued that the British should convert its excess military fleet—which had just headlined Britain's victory in the Seven Years' War—into a fleet of whale fishing ships, manned by the sailors and officers who otherwise would have been discharged. As a port town, the proposal would likely have been of interest to many in the Philadelphia community, including both sailors and merchants.

Beginning at the middle of page 2, Bradford printed news from other colonies, in reverse order of proximity to Philadelphia. First, therefore, came news from Savannah—two paragraphs pulled from the *Georgia Gazette* of April 7, 1763, followed by news from Charleston (the *South-Carolina Gazette* of April 30, 1763), Boston (May 16), Newport (the *Newport Mercury* of May 16), and New York (May 19 and May 23).[16] The last three most likely arrived from what Philadelphians would have then described as the "eastward" via the same post rider. Bradford chose these stories from a variety of possibilities in each paper, and why he chose a particular story may be difficult to discern: someone might have brought something to his attention, he might

TO BE SOLD BY
JAMES CHILD and SON,
[illegible]

LOAF Sugar at twelve pence, thirteen pence, and eighteen pence per Pound, Mollasses, Muscovado Sugar, &c. Also a neat assortment of Hosiery.

THE RAVEN,
A VERY [illegible]

TO BE SOLD,
By BRYAN BRUIN,
[illegible]
In FREDERICK viz.
O[illegible]
In HAMPSHIRE viz.
[illegible]
BRYAN BRUIN.

March 24.
Just imported and to be Sold by
JAMES CLAYPOOLE,
In Walnut Street, Philadelphia:
B[illegible]

T[illegible]

April 7.
TO BE SOLD
A Likely young Negro Wench; [illegible]
ROBERT LETTIS HOOPER, junr.

TO BE SOLD BY
DANIEL ROBERDEAU,
A[illegible]

All Persons that stand indebted for
TONNAGE and IMPORT DUTIES,
A[illegible]
ENOCH STORY Collector.

Just come to Hand, and to be Sold by
John and Peter Chevalier,
A[illegible]

A general CHART of the Coast of the Province of
LOUISSIANA,
[illegible]
MISSISSIPPI and APALACHY,
In the GULPH of MEXICO.
[illegible] MATTHEW CLARKSON.

A[illegible]

IMPORTED and to be Sold by DANIEL WILLIAMS [illegible]
Boulting Cloths.
M[illegible]

ROBERT LEVERS,
I[illegible]

March 10.
Just imported in the Ship Britannia, Capt. [illegible] from London, and to be sold by
THOMAS SMITH,
A Very large Assortment of Goods, suitable for the Season.
N. B. He once more Requests all those that are indebted to him, to Discharge their Accompts.

CHOICE [illegible]
JOHN MIFFLIN [illegible]

TO BE SOLD BY
WILLING, MORRIS and Company,
B[illegible]

Just imported in the last Vessels from London, and to be sold Wholesale and Retail by
SMITH and HARRIS,
A[illegible]

T[illegible]

Just imported and to be sold by
FRANCIS and RELFE,
A[illegible]

TO BE SOLD BY
SAMUEL HOWELL,
J[illegible]

A[illegible]

PHILADELPHIA: Printed and Sold by WILLIAM BRADFORD, at the Sign of the Bible, the Corner House of Front and Market-Streets, where Persons may be supplied with this Paper at Ten Shillings a Year.—And where Advertisements are taken in.

May. THE Numb. 1068

PENNSYLVANIA JOURNAL; AND WEEKLY ADVERTISER.

THURSDAY, May 26, 1763.

AN ESSAY ON THE GOSPEL MINISTRY,
[illegible]

Just imported [illegible]
JOSHUA FISHER & SON,
T[illegible]

THREE POUNDS REWARD.
R[illegible]
ISAAC DUSHANE.

R[illegible]

TO BE SOLD BY
JOHN GIBSON,
Jamaica Spirit, and a few Hogsheads of Choice Molasses.

TO BE SOLD BY
JOHN ISAAC REDWAR,
CHOICE [illegible]

TO BE SOLD BY
REES MEREDITH,
A[illegible]

T[illegible]
MARGARET CHURCH.

A[illegible]

SAYINGS.

T[illegible]
REESE MEREDITH,
WILLIAM FISHER,
JOSEPH FOX,
THOMAS WHARTON.

JAMES & THOMAS WALLACE,
A[illegible]

TO BE SOLD BY
Robert Lettis Hooper Junior,
T[illegible]

The Subscriber begs Leave to inform the Public,
T[illegible]

WILLIAM SYMONDS,
A LARGE [illegible]

ROBERT RITCHIE,
A[illegible]

Imported in the last Vessels from London, and Liverpool, and to be sold cheap by
JOHN COX,
A Neat assortment of European and India Goods, [illegible]

GARRETT and GEORGE MEADE,
A[illegible]

The German FLUTE
Taught by GEORGE D'EISSENBERG.
I[illegible]

ABRAHAM USHER & JOHN MALCOLM
B[illegible]

Figure 1. Pennsylvania Journal, May 26, 1763, pages 1 and 4. Historical Society of Pennsylvania

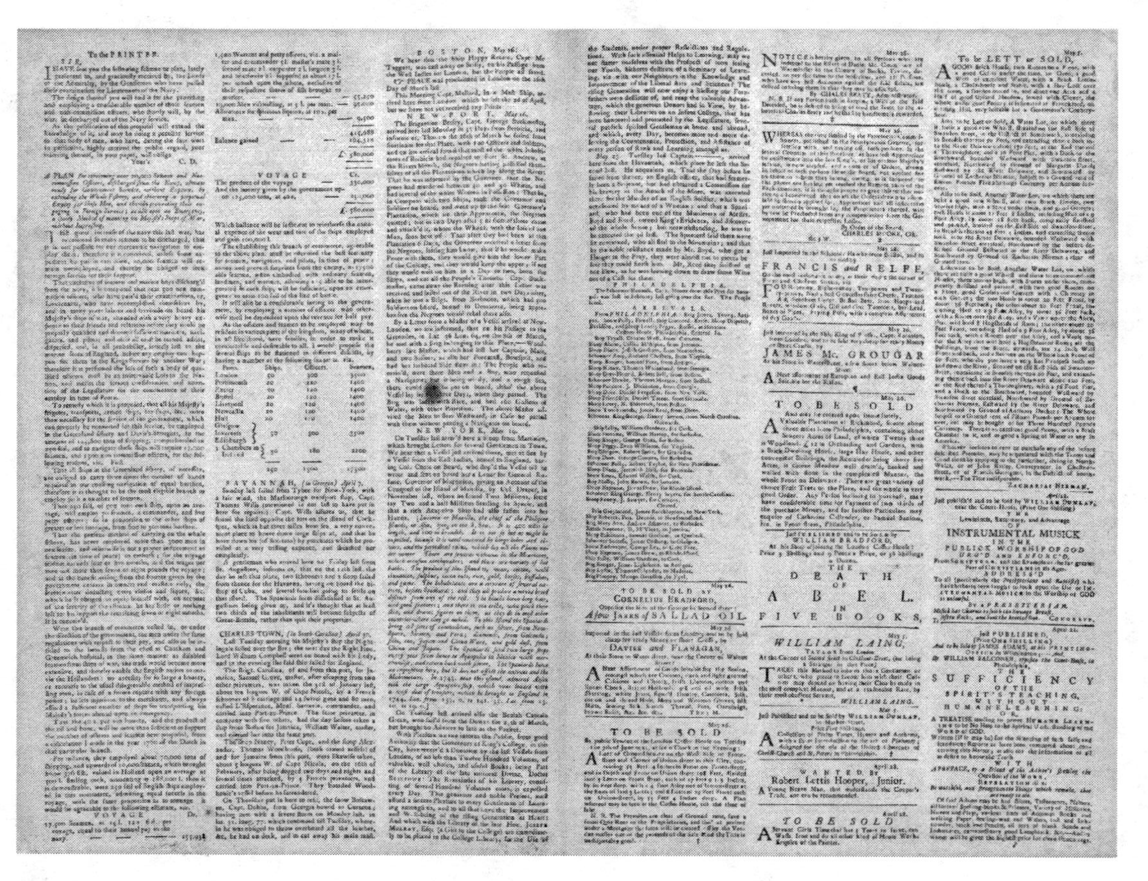

To the PRINTER.

BOSTON, May 16.

NEW-YORK, May 16.

PHILADELPHIA.

TO BE SOLD

FRANCIS and RELFE,

JAMES Mc. GROUGAR

TO BE SOLD

THE DEATH OF ABEL. IN FIVE BOOKS.

WILLIAM LAING,

TO BE SOLD

WANTED BY Robert Lettis Hooper, Junior.

TO BE SOLD

To be LETT or SOLD,

INSTRUMENTAL MUSICK

THE SUFFICIENCY OF THE SPIRIT'S TEACHING, WITHOUT HUMANE-LEARNING.

Figure 2. Pennsylvania Journal, May 26, 1763, pages 2 and 3. Historical Society of Pennsylvania

have had a commercial interest in those areas (though probably not), or these simply were the most salacious stories available that week. The nature of the newspaper meant that the printer reproduced his interests and circulated them further without ever revealing the reasons for the reproduction.

Finally, on page 3, Bradford printed the briefest of notices of Philadelphia news: a three-line paragraph about the sinking of the schooner *Hannah* in February en route from Philadelphia to Senegal. The only other local news that week—and, again, this is typical—was a list of recent ship arrivals in port, which was important for merchants trying to track their shipments of goods. For the most part, local news was overlooked in the newspapers of colonial America, and for good reason. What Philadelphians needed to *read* about was news from elsewhere. Colonial towns were exceedingly small, and all the more so when one considers how restricted was the circle of middling and elite men who dealt in business and politics and served as the intended audience of newspapers.[17]

A newspaper served several functions in the business of the printing office and the political life of the town and colony. Primarily, it generated revenue for the printer through subscriptions and advertisements. By demonstrating the printer's skill, it helped to drum up other printing business. It also served the public's interest in news—hence the moniker of the "public prints." Printers who started newspapers in the 1760s and 1770s made these and other goals explicit, typically through an essay in the newspaper's first issue that laid out the rationale for publication. William Brown and Thomas Gilmore wrote in the first issue of the *Quebec Gazette* that their newspaper would include the latest materials on Britain and the Continent, as well as the other American colonies, plus essays and poetry on virtue and morality, "*so that blending* Philosophy, *with* Politics, History, &c. *the youth of both sexes will be improved, and persons of all ranks agreeably and usefully entertained.*"[18] When he began his *New-York Gazetteer*, James Rivington of New York reasoned in a similar vein: "*The general Interests and State of Nations, the Causes of the Rise and Declension of Empires, and the singular and important Occurrences that arise in any Quarter of the Globe, when faithfully exhibited, cannot fail of interesting, entertaining, and in some Degree, instructing every Reader.*"[19] Printers tried to appeal to as many different readers as possible, asking them to support their newspapers because it would serve the need for political news, belles lettres, science, and other topics.

The small scale of newspaper production in the colonies contrasted sharply with contemporary practice in England. Many of the newspapers in London were owned by conglomerates of several booksellers and shareholders. The larger size of the newspaper industry in England meant that both government and opposition supporters financed newspaper operations, which helped nurture a partisan press. Such newspaper operations were also more complex. A single printer in London frequently operated multiple newspapers, including a weekly, a daily, and one that appeared two to three times per week. Provincial towns such as Bristol, Plymouth, York, and Newcastle had less substantial press operations during the eighteenth century but still maintained a more varied debate in print than occurred in the colonies.[20] Still, the process of news distribution and production was in other respects similar. In England, as in the colonies, newspapers included advertisements and news compiled from a variety of sources, with only a small portion of new and original text in each issue. Newspaper transport and delivery relied on the use of local boys within town and on the postal system for distribution beyond the metropole.

In the colonies, newspapers developed more slowly than in England and under less competitive market conditions. The first newspaper published in the colonies, *Publick Occurrences*, famously produced only one issue. The Massachusetts government censored Benjamin Harris, the publisher of that newspaper, because officials believed the publication of any news would harm the government.[21] After 1704, however, a succession of printers began newspapers in the major port towns of Boston, New York, and Philadelphia. These small, weekly newspapers largely published advertisements and imperial news, along with the occasional essay or literary work.[22] Newspaper printers aimed to create a running historical journal of imperial events rather than to shape public debate. Their print runs were relatively tiny, usually a few hundred issues per week. Because the market for newspapers was so small, very few towns could sustain a second newspaper (and even fewer a third). By the end of the Seven Years' War, this situation began to change. Newspapers had spread out of the major port towns into secondary towns and southern ports, and several towns had multiple news outlets.

This study works from the premise that communication mattered, not just the existence of texts but the processes that led to their creation and circulation around the colonies and the Atlantic world. In the past decade, numerous historians have attested to the importance of networks in shaping political, social, and cultural discourse throughout the various empires engaged across the Atlantic Ocean. For instance, scholars have mapped the networks of elite intellectuals in both Europe and the Americas in order to understand the creation of knowledge (especially in the areas of medicine and science), interactions between Native Americans and Europeans, politics and diplomacy, and a range of other areas.[23] The emergence of digital history and the digital humanities also means that we can now visualize more clearly than ever before the networks of key intellectual figures, from early modern scientists such as Francis Bacon to European and American Enlightenment figures such as Voltaire, David Hume, and even Benjamin Franklin.[24] Historians have been able to make these networks visible because of the voluminous correspondence available for these groups of individuals. By contrast, the networks I examine are much more difficult to map precisely. Because the printers in this study did not leave a clear documentary record tracing their sources (Franklin and his associates being the notable exception), their networks must be examined and reconstructed indirectly through their sharing of news items in print media, references among printers to one another, and occasional manuscript records such as letters or account books.

Print has long been seen as a central venue for understanding the debates surrounding the American Revolution, in particular by political historians. Over the past fifty years, dozens of historians have argued either explicitly or implicitly for the importance of texts to understanding the Revolution, and especially particular genres of print, most notably pamphlets and newspapers.[25] These historians have stipulated that some of the power of the arguments contained in the texts they examine came from their status as printed objects that circulated throughout the colonies (and sometimes across the Atlantic Ocean). However, many scholars of the Revolution have treated these texts only for the words written on the page without particular concern for the circumstances of their production or circulation. This holds true even for those scholars who have explicitly invoked the circulation of print as part of their arguments. It occurs, but largely in the background, and without the active intervention of any people except for the authors.[26]

Attending to the life of texts beyond the words they contain requires more sustained attention to the material processes through which texts first came into being and then reached the eyes of readers. The history of the book, as the field is known, offers an interdisciplinary method to examine the production and circulation of texts as a material process.[27] In so doing, these scholars have opened up new lines of inquiry that invite consideration of texts not as complete and total but rather as constructed and contingent.[28] Using the approach of book history, this volume opens up the processes that occurred inside printing shops to reveal how the work of printing, beginning with the seemingly mundane mechanical tasks of setting type and pulling the press, shaped the politics of the American Revolution.

Because the book situates its argument at the intersection of print and politics, it also engages with two theoretical models that seek to explain the operation of political debate in public and the development of nationalism. Jürgen Habermas, writing in the early 1960s, suggested that printing was central to the creation of a sphere of political discourse distinct from and in opposition to the state. Print, he argued, made it possible for bourgeois men to engage in rational political debate uninfluenced by the social position of the participants. That debate occurred not only in the metaphorical space of the newspaper but also in conversations in coffeehouses, taverns, salons, and post offices. Because Habermas argued that eighteenth-century commercial life encouraged the activities and institutions that made the formation of a public sphere possible, historians and other scholars since the translation of his work in the late 1980s have taken a great interest in identifying the specific kinds of places that were hospitable to political debate and the formation of public opinion, including the coffeehouse, the tavern, and the salon.[29] Despite his emphasis on print as a mechanism for public debate, however, printers themselves vanish from his account, serving merely as neutral instruments of the public sphere without influence of their own. In his classic work that applied Habermasian theory to eighteenth-century British North America, literary scholar Michael Warner linked the public sphere to republican political ideology. The two, he suggested, in fact mutually reinforced one another in print through their emphasis on disinterested and autonomous civic participation.[30] The present study operates at the hinge of Warner's analysis to explore how printers, through their control of the mechanisms of print, shaped political texts and debate, helped to produce coordinated resistance to imperial politics, and began to develop a national communications infrastructure. In short, printers promoted sentiments that could fairly be called protonationalist and fashioned a public sphere that was deeply infused with commercial interests throughout.

In this respect, Benedict Anderson, with his explicit emphasis on "print capitalism" as the underpinning of national identity is a better guide to the role of print and newspapers in the politics of the Revolution than Habermas, who tended to view economic interest as contaminating and degrading political debate. "Print capitalism"—that is, the drive to make a profit through the circulation of news and information in print—was for Anderson the key factor in the development of nationalism. The ideology of the nation, he argues, developed out of the "imagined community," the sense of belonging to a common group that grew among people who shared a common language, administrative structure, and identity.[31] The common consumption of print, especially newspapers, fostered this sense of belonging. In fact, Anderson writes, "the printer's office emerged as the key to North American communications and community intellectual life."[32] How that process occurred, however, Anderson leaves unexamined.[33]

The straightforward narrative of participation in a public sphere and the creation of national consciousness turns out to be more complicated in Revolutionary America than the theorists implied. Trish Loughran, for instance, has argued that texts such as *Common Sense* and *The Federalist* projected national unity in order to create unity that did not exist in reality. They therefore obscured just how decentralized American print networks were until just before the Civil War. Nationalism, in her telling, was thus idealized far more than it was realized during the Revolutionary era.[34] In a similar vein, Robert Parkinson has shown how Patriot leaders circulated appeals to a "common cause" through the press in order to unite white Americans against the British during the Revolutionary War. For Parkinson, the "common cause"—which highlighted whiteness in particular as a core shared characteristic of Americanness—served far more than simply a rhetorical function but instead became the means (circulated through the press) to activate the American population to fight against the British and their Native allies.[35] These scholars are only the most prominent among several others who have enlivened the concept of the "imagined community" by expanding our view of how Americans considered and debated the idea of the nation in the early decades of the United States.[36]

Revolutionary Networks offers a path to bring together these disparate approaches through the business practices of the printing trade. The members of the trade, all artisans trained via medieval custom in "the Art and Mistery [sic] of a Printer,"[37] shared specific knowledge about how to construct and edit a newspaper, publish a pamphlet, and work to circulate and spread the information that came through the hubs that were their print shops. In addition, they learned practices that encouraged them to share information across long distances, whether between two colonial ports or from a colonial port to London or other cities in Europe. During the American Revolution, these printers combined these two sets of knowledge—the small-scale practice of editing, collecting, and reprinting within their offices, and the broader-scale commercial practice of circulating print and information—to create a set of networks that generated publicly acceptable accounts of anti-imperial resistance that in turn facilitated further debate and contest.

It is precisely the commercial concerns that have long been set to the side that prefigured and eventually shaped what political rhetoric was put before the public and how it was framed. Historians have often granted credence to the idea that the American Revolution was economic in nature. In this volume, I complicate that supposition by arguing that the economic concerns of printers—often overlooked as a matter of policy—nonetheless had a profound impact on the construction and circulation of political arguments as well as on the reception of those arguments and the subsequent generation of public opinion. These commercial interests further shaped the communications infrastructure of both Britain's colonies and the new United States.[38]

The book moves chronologically through the era of the American Revolution in order to show the development of the networks of printers and how they contributed to the process of creating first a revolution and then the new nation. By about 1763, more than fifty printers were active in British North America, and they had begun to develop a set of commercial connections that in some cases amount to what we might call a network. Chapter 1 outlines this set of connections and explores the world that printers inhabited just before the imperial crisis, one in which they existed in a middling but never central position. Much of their work was manual labor, but success depended on their literacy and artistic skill. Although they were tradesmen, they interacted with the elites of society.

Chapter 2 examines how printers shaped and refined political debate during the imperial crisis itself. In particular, it argues that we must reexamine the Stamp Act crisis—the classic "opening act" of the Revolutionary drama—as an event made by printers. As the targets of the tax, printers were crucially invested in preventing the act's effects regardless of their position on the question of Parliament's right to pass such a law. The protests that swarmed the streets of dozens of towns gained power and force to influence public opinion as printers reprinted and circulated news stories about them, and printers were assiduous in portraying the threat of the Stamp Act as dire to the colonies, to their newspapers, and to the freedom of the press.

Following the success of the Stamp Act protests, printers continued to play a key role in protesting against imperial policies from the Townshend Acts of 1767 to the Tea Act of 1773. Chapters 3 and 4 offer a particular focus on how printers amplified arguments such as those in the letters of the Pennsylvania Farmer through the process of circulation. By 1773, the networks that printers had developed for their businesses and then mobilized for political purposes had become extraordinarily effective in large part because printers had begun to coordinate their activities with extralegal protest groups like the Boston Committee of Correspondence and the Sons of Liberty. In so doing, they became even more effective at utilizing their networks for protest against the Tea Act and not only moved public opinion but also shaped the response in cities along the Atlantic coast with the effective circulation of news. In the following years, printers moved to overthrow the British imperial post office to further cement the security of their information network, and after the beginning of the Revolutionary War they played a key role in circulating news about battles and promoting political propaganda such as *Common Sense*.

The war, however, would prove difficult for printers and their businesses. As the final two chapters show, the political and military difficulties that accompanied the war wreaked havoc on the ability of printers to do their jobs in the way they had come to do them. Many had to scatter from their homes as one army or the other (or both) took possession of towns along the Atlantic coast. When not forced to relocate, printers suddenly had difficulty getting access to all of the tools of their trade: paper and ink were in scant supply, it was hard to replace worn out type and press parts, and information flows were disrupted by armies and navies. When they could produce a newspaper or pamphlet, few residents could afford to subscribe or pay for them. The spread of war thus brought extreme clarity to the political interests of printers. That is, simply by keeping their shops open in particular towns, printers revealed their allegiances. In an ironic twist, that meant that printers thought far less about how to navigate politics in their publications and far more about the economic concerns that plagued their ability to stay in business.

In the war's aftermath, citizens in fourteen states (including Vermont) debated how and when the United States should act as a single nation and began the long process of piecing together their lives. For members of the printing trade, the 1780s were a trying time. Many had lost their livelihoods entirely during the war, never to recover them. Others returned to their old homes or established themselves in new ones and continued on in business. As they did so, the trade bifurcated. An increasing number set out toward the North American interior with their compatriots, beginning a marked shift in focus for the printing trade from the Atlantic coast to the interior of the United States. These printers continued to run their offices using colonial business strategies: they appealed to readers across the political spectrum, avowed their own impartiality, and relied on government contracts and other subsidies to operate. In Atlantic cities, printers competed for readers with openly political publications and practices new to American markets, such as daily papers. By the end of the decade well over four hundred printers were active in the United States, nearly eight times as many as in 1763. Because of their focus on long-distance communication, and with travel often necessary to take up a position as a master printer, many of the participants in the printing trade were more supportive of a national identity and unified government than the average American. As such, printers largely promoted the federalist pro-Constitution cause, setting them up to become key participants in the national public sphere of the new federal government.

The Business and Economic World of the Late Colonial Printing Trade

Although direct accounts of a typical day in a colonial printing office are few, circumstantial evidence affords a relatively reliable picture. Imagine yourself standing along the Parade in Newport, Rhode Island, on a cool autumn day in October 1763. Just a few streets away, you can hear the sounds from Long Wharf as dock workers make ropes, load a ship with goods bound for the Caribbean, and perhaps unload a ship recently arrived from Liverpool, replete with fine European products and the latest news from the Continent. Crossing the street, you see the Franklin printing house, indicating that you have arrived at the shop of Samuel Hall, at one time the partner of Ann Franklin (sister-in-law of Benjamin) and now master in his own right after her death earlier that year.[1]

As you approach the doorway, you see a flurry of activity inside. In one corner, a journeyman sits with his composing stick, furrowing his brow to set the type for a page of *An Astronomical Diary: Or, Almanack* by Nathaniel Ames, a noted New England polymath. Hall was publishing an almanac for the first time, so he simply reprinted Ames's work as it had been calculated for Boston, some seventy miles to the north.[2] Two teenage apprentices are cleaning in another corner, and at the center of the room, where one of the two printing presses in the office stands, you see two men working feverishly to run the pages of that week's *Newport Mercury* through the press so it can dry in time to print the other side tomorrow. At the counter at the front, you see a young woman, either Hall's wife or perhaps an employee, open the account book to mark a debt paid from a local merchant. And next to her you see Hall himself, wearing an ink-stained apron but standing with all the pride a twenty-three-year-old master printer can muster. In front of him the October 13 edition of the *Pennsylvania Gazette* lays open, a pen at its side to select paragraphs to reprint.[3] He has cleaned his hands of ink as best he could to speak to Newport's customs officer, who has just delivered the list of ships that have recently arrived and are planning to depart. As part of Hall's obligation to serve the public, the list will go straight to the compositor working on that week's *Mercury*.

The scene described is a composite but was typical of the process replicated hundreds of times each year in printing offices throughout British North America. By the end of the Seven Years' War, printers plied their trade throughout Britain's Atlantic colonies, intersecting with one another through kinship and professional relationships. Through their everyday commercial activities, printers laid the foundation for a series of interlocking networks that spanned nearly the entirety of Britain's colonies in the New World, including not only the thirteen colonies but also Canada and the islands of the West Indies. Printers often traversed the networks themselves, occasionally working in three or more colonial towns during their careers. As they worked and moved as apprentices, journeymen, and master printers, they encountered colleagues who extended their personal contacts across the map of Britain's possessions. These connections proved crucial to printers, who relied on their extended network to sustain their businesses: the contacts were vital for the sharing of resources, in particular information. During the British imperial crisis from 1763 to 1776, these networks would begin to crystallize into a new American infrastructure of political communications.

The networks of colonial printers, which began to develop in the late seventeenth century, were foremost about effective business practices. In order to sustain a printing office, a printer needed reliable and consistent sources of information and effective channels to circulate and distribute his products. By the 1760s, some networks had matured to sufficient size and geographic breadth to give the printers who participated in them significant leverage in circulating political news throughout colonial North America and the Atlantic world. This chapter explores how printers worked and connected just as the Seven Years' War ended, with the imperial crisis just beyond the visible horizon. Because printers were artisans, they were in many respects similar to others in the skilled handicraft trades. At the local level, printers faced substantial economic obstacles as they worked to finance their operations and develop long-lasting and lucrative businesses. However, the geographically expansive nature of printing stimulated the creation of trade networks that more closely resembled those of merchants. Printers' success or failure depended in no small part on their ability to make connections with others in the trade and with outside parties at home and throughout the Atlantic world. These relationships were crucial for printers, who needed them to gain access to information networks on both a local and an Atlantic scale. An exploration of these elements of printers' professional lives reveals the contours of printers' business world as it stood in the early 1760s, just before the imperial crisis shook the foundations of colonial society.

Printers as Artisans

To describe printers as a group is difficult because they had such varying socioeconomic characteristics. A few printers met with economic success of a relatively definable sort and came to be viewed as among the leading citizens in their towns. They were their own men, politically and economically. Another small group of printers included, as the cliché went, "meer mechanics."[4] These men were not very well educated in the liberal sense and lacked the skill in refined manners that would have allowed them to rise in society. To the extent that any of these printers gained notoriety, it was as the tool of more powerful forces, whether a rich merchant or an influential political faction (John Peter Zenger is probably the epitome of this group). Most printers, however, fell into neither of these extremes. They muddled along, perhaps having a flush year or two where they published a popular tract or had their newspaper gain readership, but mostly just eking out an existence, not unlike most of their neighbors. But three of their characteristics can be described in common. First, they were decidedly in the mushy middle of society—or, to put it less charitably, they often found themselves standing directly over the fault lines of society. Second, they operated their businesses along conventional eighteenth-century lines, with the family as the model and the basic unit of economic life. Third—and here they differed from many of their artisan brethren—they were decidedly mobile, both within the North American colonies and across the Atlantic Ocean.

As artisans, printers were highly skilled manual laborers. Printing required a sophisticated labor process to produce commodities for sale; the lengthy apprenticeship period (ideally about seven years) was not wasted. At the same time they were not among the social elite of artisans, those who worked with luxury goods like gold and silver.[5] They were for the most part literate. That is, they had to be both capable readers and at the same time aware of trends in the arts, sciences, politics, the law, and other areas of culture. As voracious consumers of information (in order to become vociferous distributors of it), printers interacted frequently with people across a wide range of the socioeconomic spectrum, from mariners carrying news from abroad to wealthy merchants wishing to advertise their ships' goods.[6] Under ideal circumstances, printers ranked above many other artisans, sailors, and common laborers because of their skill sets and social interactions. But conversing and conducting business with elites rarely enabled printers to join their ranks. Printing necessitated intense manual labor, and printers clearly bore the physical marks of their work—ink-stained hands, muscle aches and pains, and a stooped posture as years of physical labor took their toll.

Printers organized their shops around the family unit, and the master printer served as head of both the business and the family, though he would not have recognized a distinction between the two.[7] His wife was crucial to the successful operation of a printing office not only as a marriage partner but also as a skilled participant in the household economy. In addition to the gendered labor of feeding, clothing, and caring for her family, as well as apprentices and journeymen (who slept under the same roof), she may have kept the account books or operated a related segment of the business, such as retailing stationery or other goods. Benjamin Franklin, for example, noted in his autobiography that his wife Deborah "assisted me cheerfully in my Business, folding and stitching Pamphlets, tending Shop, purchasing old Linen Rags for the Paper-makers, etc. etc."[8]

Printing offices usually employed several other workers. The most significant of these were the apprentices and journeymen, young men training in hopes of becoming master printers themselves some day. They undertook a range of tasks, from the new apprentice who might spend most of his time cleaning and running errands to a shop foreman who served as a trusted aide of the master. Unmarried women also pursued work on their own in printing offices. In 1773 Ebenezer Hazard suggested to William Bradford in Philadelphia that he should hire his sister Betsey, "a smart Girl" that he thought "would suit [Bradford] very well" for running his shop.[9] Many master printers utilized enslaved labor as well, whose tasks ranged along the same spectrum as apprentices and journeymen without the prospect of advancement. Records on their careers are scant. We know a fair amount about Peter Fleet, for example, who worked for Thomas Fleet, printer of the *Boston Evening-Post*, in the 1740s. Though enslaved, Peter Fleet drafted a will in which he identified himself as a printer.[10] Numerous other master printers held slaves, including southerners like Robert Wells as well as Benjamin Franklin.[11]

The family served as a key conduit for young men to enter the trade. As part of their indenture agreements, apprentices were fed and clothed as members of the household unit. Most young printers, like their counterparts in other artisanal trades, entered into apprenticeships because of relatives who were already at work as printers (most often a father or uncle). Of the fifty-nine printers active in 1763, for example, at least thirty-six had a family connection within the trade. Such intergenerational kinship ties helped to transmit skills and provided a built-in support system. These young printers had the advantage of working with a set of business connections already established by their relatives and their relatives' partners. Yet even apprentices who lacked explicit kinship ties nonetheless became part of their master's family, which could lead to some of the same benefits.

The connections of kinship over time afforded a few families significant control of the printing trade, particularly in New England. The most influential of these families was the Green clan in New England, founded by Samuel Green, who operated a press in Cambridge, Massachusetts, in the mid-seventeenth century. Between 1700 and 1750, more than half of New England printers were Greens.[12] By the imperial crisis, more than forty printers could trace at least part of their professional lineage to a Green connection, either as a descendant or because they worked for or with a member of the family. Blood descendants operated printing offices throughout New England and the mid-Atlantic colonies, from Boston in Massachusetts to Annapolis in Maryland (map 1). In addition, many other printers had links to one or more members of the Green family. Benjamin Edes and John Gill, for example, apprenticed in Boston with Samuel Kneeland, whose mother Mary was a Green; Kneeland, in turn, apprenticed with his uncle Bartholomew Green. Edes and Gill opened a joint venture in 1755 and assumed publication of Kneeland's newspaper, the *Boston Gazette*.[13]

Other families also developed lineages of some breadth. William Bradford served as the first printer in New York City in the late seventeenth century. He initiated a dynasty that spread to Philadelphia and included several printers, among whom were Andrew Bradford, Andrew's widow Cornelia, and his namesake nephew William Bradford, who was active during the Revolutionary period. In New England, the Fowle family numbered some eleven printers in Boston and Portsmouth, New Hampshire, during the imperial crisis. Even non-English printers fostered familial connections: the Sower family, German-language printers in Germantown, Pennsylvania, trained father to son, though they were constrained from expanding because they served a smaller linguistic community. These linkages could not guarantee success, but they offered a somewhat more well-trod and stable path into the trade.

When such familial connections did not exist by birth, those in the print-

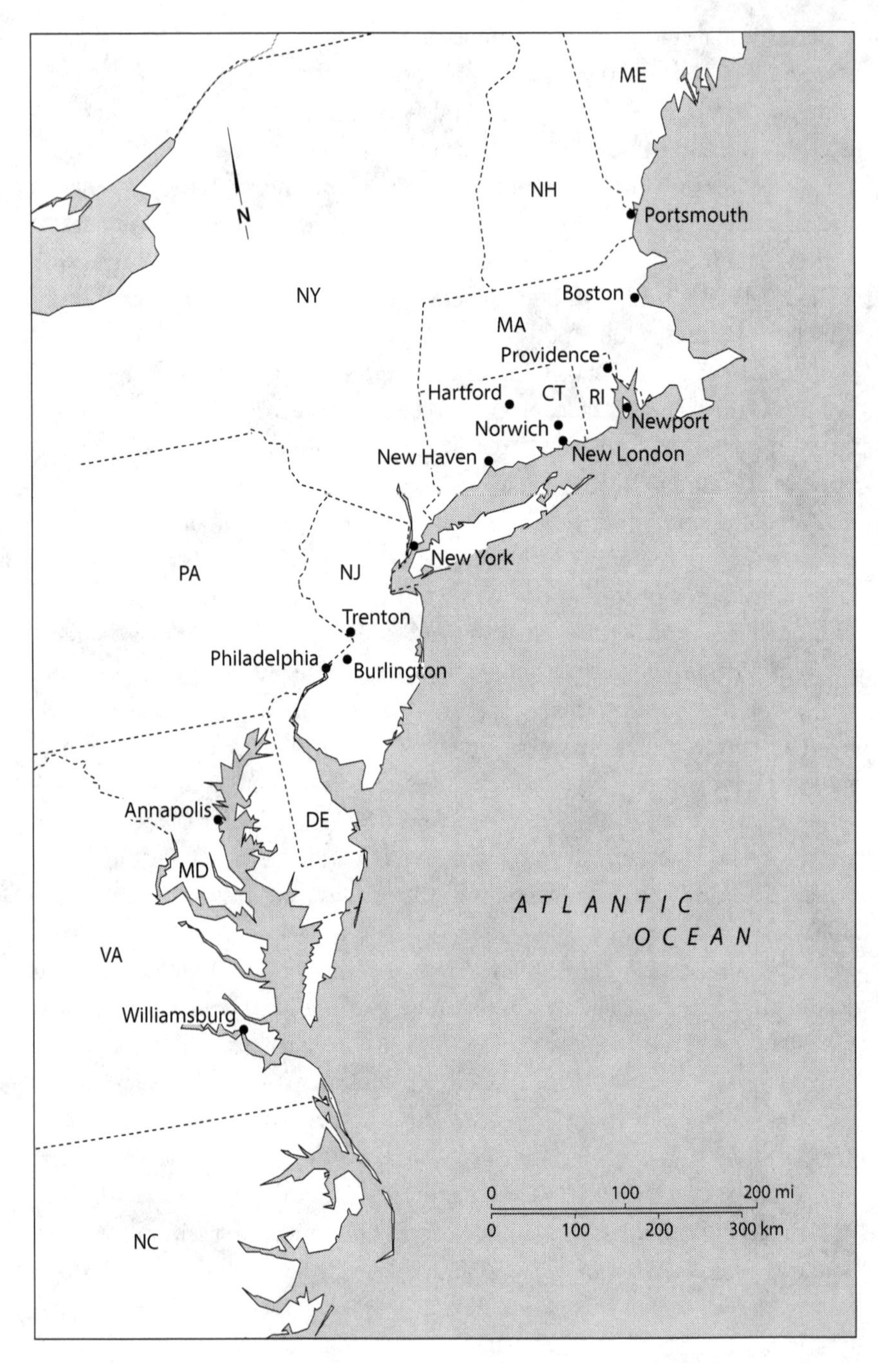

Map 1. Locations of printing offices run by members of the Green family, ca. 1763–76

ing trade could encourage them through marriage alliances. As with other artisanal trades, printing apprentices often found their future wives under the same roof.[14] Of the master printers active during the Revolutionary era, 414 are known to have married; of these, at least 45 married someone from a printing family.[15] Three factors keep that number from being higher: records were rarely clear about all but the most obvious connections; many could not by law marry the master's daughter, as she might very well also be the apprentice's sister or cousin; and only one apprentice could marry each female relative, and a master printer could marry off only as many daughters as he had. Not one but two Virginia printers, for example, married Rosanna Hunter, the half-sister of Williamsburg printer William Hunter. Joseph Royle married Rosanna first. He had worked as the foreman in Hunter's office and then, after Hunter's death in 1761, operated the office on behalf of his infant son, William, Jr. After Royle died in 1766, Rosanna married John Dixon, who took over the office and, with it, publication of the *Virginia Gazette*.

For a few women, family connections provided a window of opportunity in which they could assert themselves in public as commercial and political actors. Through the colonial period, about fifteen women also became either printers or proprietors of printing offices. Most frequently, they did so as a result of their husbands' deaths.[16] In these cases widows usually served as the executors of their deceased husbands' estates. Their management of the office was usually intended to be temporary while they addressed their deceased husbands' accounts.[17] Most often, these women turned over everyday operations of the shop to a son, the shop foreman, or a favored apprentice within a few years. As part of the arrangement, the new proprietor typically agreed to provide for the widow a portion of the office's revenue. This was true, for example, of William Goddard's mother, Sarah. After he left Providence in 1765 to pursue opportunities in New York, she ran the Providence office under the imprint of "Sarah Goddard and Company" for two years in tandem with John Carter, a one-time apprentice and journeyman. By 1767, however, she decided she was too old to continue, sold the shop to Carter, and moved to Philadelphia, where her son had opened a new printing office.[18] Even there she did not retire but continued to oversee the operations of her son's business.

During the Revolutionary era, several women went beyond this caretaker role and pursued independent careers. All of these women continued the newspapers that their husbands had operated and attempted to maintain the intercolonial contacts essential for gathering news. At least five widows—Margaret Draper of Boston, Clementina Rind of Williamsburg, Anne Catharine Green of Annapolis, Elizabeth Holt of New York, and Ann Timothy of Charleston—assumed not only their husband's printing offices but also their lucrative appointments as official printers to their respective colonies and states. Green would remain printer for the Maryland Assembly until her death in 1775 and even commissioned a portrait by Charles Willson Peale (fig. 3). The portrait, a mark of gentility, displayed her in the modest clothing of a tradeswoman holding a copy of the *Maryland Gazette*, her newspaper. Such a gesture was extraordinarily rare for a portrait of a woman in the eighteenth century and set her apart from her peers in her gender and in her profession. Very few portraits of women showcased their nondomestic craft skills, but a few other printers became as successful as Green. Mary Katherine Goddard was perhaps the most distinctive of these women. The sister of printer William Goddard, she took over his printing office in Baltimore while he was away working for the Continental Post Office. She ran the printing office and published the *Maryland Journal* for several decades. Eventually, Goddard was appointed the first postmistress of Baltimore by the Continental Congress, a position in which she served until 1789.[19] Though considerably fewer in number and lacking access to formal apprenticeships for training, the women who entered the trade brought considerable expertise and training in printing specifically and business practices more generally.

The contingent nature of the printing trade meant that printers were of necessity highly mobile. Many towns could support only one printer (if that), so an apprentice or journeyman who wanted to open his own office typically had to move to a new town, sometimes at a great distance.[20] As a result, printers were among the most mobile occupational groups in the colonies, perhaps second only to sailors and mariners in sojourning across numerous regions of North America and the Caribbean. Frequent travel also meant that printers could develop a broad range of contacts before they ever opened offices. William Brown, for instance, built up contacts based in Philadelphia that helped him develop a printing career in Canada. Born in Scotland in the late 1730s, Brown came to America with his family.[21] For a short time he attended the College of William and Mary before he was apprenticed as a printer in Philadelphia to William Dunlap, himself an Irish immigrant. Once Brown reached his legal majority, Dunlap arranged to send him to Barbados to open a printing office there. After three years, Brown returned because of poor health and "decided to go live in a more hospitable climate."[22] He chose Quebec, only recently part of the British empire, and made a partnership agreement with Thomas Gilmore in which each put up £72. Dunlap also offered a guarantee of £150

Figure 3. Charles Willson Peale, *Anne Catharine Hoof Green*, 1769. Oil on canvas. National Portrait Gallery, Smithsonian Institution; partial gift with funding from the Smithsonian Collections Acquisitions Program and the Governor's Mansion Foundation of Maryland

to support the endeavor. One scholar has posited that Benjamin Franklin knew about the arrangement because he had sent Dunlap to scout the possibility of a post office in Quebec. The proposition is further supported because Gilmore went to London to acquire type from William Caslon, one of the leading type founders in England, and other materials from Kendrick Peck. Brown, meanwhile, traveled to Quebec with a broadside in hand to solicit subscriptions for the *Quebec Gazette*, a bilingual newspaper that he and Gilmore began publishing in June 1764.[23] Brown was enormously successful, amassing a fortune of about £10,000 at his death in 1789, thanks in part to the connections he built as an apprentice and journeyman.

Many printers thus ended up as adults someplace other than where they had grown up and trained, and some traveled not just within North America but across the Atlantic.[24] Of the 475 printers with known birthplaces, 116 are known to have immigrated, or about 24 percent. Over time, the number of immigrant printers increased, from 18 active in 1756–60 to 75 active in 1791–95.[25] The pace of immigration also increased during the period, though with a small sample it does not appear to have increased a great deal. We know, however, that some new printers came to North America nearly every year between the 1750s and 1790s.[26]

As the number of immigrants grew, the list of places from which they came shifted, largely in proportion to the origins of the broader immigrant population. Most came from England, Scotland, and Ireland, but a significant minority hailed from Germany; smaller numbers came from France, Holland, Switzerland, and even Russia.[27] In the late 1750s, more than one-third of foreign-born printers were German, who published largely in Philadelphia and its hinterlands, where the preponderance of the German population lived. In the 1750s and 1760s, Pennsylvania boasted several German-language newspapers, including the *Philadelphische Zeitung* (which changed its name to the *Germantowner Zeitung* in 1762), published by Anthony Armbruster; the *Pensylvanische Berichte*, published in Germantown by Christopher Sower Jr. and his son, Christopher Sower 3rd; and the *Wöchentliche Philadelphische Staatsbote*, published by John Henry Miller. After the war, however, the number of German immigrant printers declined as the flow of German immigrants slowed.[28] The German printers active in the 1780s, therefore, were primarily the veterans of that earlier generation of migration. Several Armbruster and Sower apprentices also entered the trade, but by the early 1790s Germans made up barely an eighth of the immigrant printers in the United States.

By contrast, the proportion of immigrants from the British Isles (including Ireland) rose steadily through the second half of the eighteenth century. Where in the late 1750s British and Irish printers represented about half of the immigrant printers in the colonies, by the early 1790s they accounted for nearly 80 percent—most of them English or Irish, with a steady and significant inflow of Scottish printers.[29] Not surprisingly, there was an influx of printers just after the end of the Revolutionary War; fifteen printers arrived between 1783 and 1785, including such important printers and publishers in the early United States as Mathew Carey, Samuel Campbell, Thomas Allen, and Thomas Dobson, all of whom came from Ireland or Scotland. On their arrival in North America, most immigrant printers stayed along the Atlantic coast. By far their most common destination, and the site of their first work as printers, was the mid-Atlantic region, where sixty-eight had their first opportunity to run an office. Philadelphia was the most popular single destination, with forty printers; New York City had thirteen. They were therefore far less likely than native-born printers to work during this period in new interior territories such as Vermont, Kentucky, Tennessee, and Ohio—in fact, only four are known to have done so, and they migrated west in large part on the basis of connections they made once they arrived.

Those printers who immigrated to North America were part of a large group of artisans and tradesmen making the transatlantic voyage, most of whom arrived owing labor or some other debt.[30] However, because of limited information about the circumstances of immigration for most printers, it is unclear under what status they traveled across the Atlantic. Nonetheless, most printers in this era came over with or for the purpose of acquiring trade skills, which put them in a better position to make headway once they reached North America. Bernard Bailyn, who tracked thousands of English and Scottish immigrants to North America during the 1760s and 1770s, found that more than half were artisans or tradesmen in roughly the same social stratum and occupational fields that printers occupied.[31] They were, Bailyn noted, a group that saw emigration "not so much a desperate escape as an opportunity to be reached for."[32] Printers in this regard were no different from their fellow travelers.

After the Revolution, immigrants composed a smaller proportion of the printing trade as the number of American-born printers skyrocketed. New entrants flooded the printing trade after the Revolution for two reasons. First, Euro-Americans and immigrants alike began to push westward into and across the Appalachian Mountains. As new towns sprung up, they rapidly sought to acquire printing operations to publish necessary forms, almanacs, and newspapers and to bring books, newspapers, magazines, and other publications from larger towns.[33] Second, the economic barriers to entry into the trade decreased rapidly after the war, in particular with the growth of American manufacturing and the ability to produce type and printing presses in the United States.[34] At the same time, one should not be fooled into thinking that every printer who entered the trade in the 1780s and 1790s was a master craftsman. The quality of the printing for many—not least those whose printing careers should be numbered in months rather than years—was often minimal.

There was one major limit to printer mobility. Both European- and American-born printers traveled extensively, but very few relocated east across the Atlantic once they had become established in North America or the West Indies. Those who published newspapers did their best to make the transoceanic connections that were vital to gathering the latest news from London. Some simply subscribed to papers like the *London Gazette* to get the "freshest advices," but others engaged agents in London to gather a variety of newspapers and magazines to send to them by the monthly packet ships. The Philadelphia printer David Hall had perhaps the best transatlantic connection. His agent in London was his good friend William Strahan; the two had apprenticed together in the Edinburgh printing office of Mosman and Brown before Hall migrated to America to work for Franklin. Strahan remained behind in London where he operated an enormously successful printing office and served in several government positions during the 1770s and 1780s.[35] Hall and Strahan were very close friends: Strahan addressed his letters "Dear Davie," and Hall named one of his sons William. Strahan's letters were enormously useful to Hall as sources of political news for his newspaper, the *Pennsylvania Gazette*. These and other connections developed out of both standard artisanal networks and the special requirements of printers to cultivate sources of information.

The Economics of Printing

Opening a printing office was an expensive undertaking. It required a large capital investment, the routine use of limited and expensive resources, and a high level of coordination with other printers and related actors. The average printing office in colonial North America required materials valued at just over £100, including a press, several sets of type, cases, composing sticks, imposing stones, frames, and other materials.[36] When Williamsburg printer William Rind died in 1773, for instance, his estate inventory cataloged about

£123 in printing equipment and materials. His two presses were not the most valuable pieces of equipment (at just £25 total); rather, it was the metal type, which was valued at more than £65, or about half the value of the entire office.[37] The largest printing office in North America, that of Benjamin Franklin and David Hall in Philadelphia, was valued at more than £300 at the dissolution of the partnership in 1766, including type worth more than £218.[38] A printer acquired these materials in various ways, whether through a partnership with a more senior printer, as an inheritance, via loans, or through the purchase of used equipment.

Because young printers lacked significant amounts of cash, they relied heavily on the system of credit and debt that tied commercial men and women to one another around the Atlantic world.[39] Because of a shortage of hard currency, or specie, a broad range of economic transactions in eighteenth-century America occurred through a complicated interplay of mutual debt and credit.[40] This informal system developed to overcome the shortage of specie and allowed printers to operate on a daily basis. In general, a businessman had to prove that he was worthy of the risk both personally and financially in order to obtain credit and favorable terms for repayment. That proof came primarily from personal connections who could serve as intermediaries to vouch for the borrower to lenders who did not know him. In a time before credit scores and background checks, a personal seal of approval was the best way to gain trust between individuals. For printers, the credit system was a double-edged sword. Merchants and other prominent community members granted them the capital to open offices and sustain them, usually in the form of book debt, promissory notes, or bills of exchange. They nonetheless had difficulty on occasion making repayments on their debts because subscribers to their newspapers and other publications made scant effort to pay off their accounts.

In order to raise capital to establish a printing office, printers turned first to their closest associates and connections. If a printer was working side by side with his father and took over the office, he simply inherited the materials, but more often additional financing was necessary. It was helpful when new printers came from well-to-do families. For example, William Goddard opened his first printing shop in Providence with enough money from his mother, Sarah Updike Goddard, to buy £120 of type from Thomas Green of New Haven.[41] A former master seeking to enhance his own prospects could also be a source of funding. Benjamin Franklin was especially assiduous in this respect. He signed partnership agreements with several of his former apprentices over the course of half a century, sending them to open offices across British North America. Well-placed benefactors could also play important roles in funding printers. It has long been suspected, for example, that Isaiah Thomas's operation, including the *Massachusetts Spy*, could not have stayed afloat for long without a loan from John Hancock, one of the richest men in Boston. Thomas himself insinuated in a letter to his onetime master and former partner Zechariah Fowle in 1772 that Hancock had put forward money to purchase types for the office.[42] These agreements were financial investments for the senior printers and merchants, and for younger printers they provided the cash flow necessary to open and operate offices before they were generating sufficient sales revenue. Such arrangements, especially in the case of those with a former master, could also provide equipment to start a business, including a printing press and sets of type.

Once fully capitalized, a printing office could begin operations. In colonial America, most shops looked very similar to one another. One master printer or two partners typically oversaw a printing office. Shops varied in size, but before the Revolution none operated more than three presses. Depending on its size, an office could employ one or more journeymen and apprentices, at least some of whom might be kinsmen of the master. Employees worked in two-man teams, which could produce a token, or 240 printed sheets, in an hour and a total of eight to ten tokens per day.[43] Many printing offices also sold other goods, in particular related items such as books or stationery, and some served as post offices.

Working six days a week, the laborers in a printing office produced a variety of printed materials. First, printers contracted for job printing, that is, printing documents for hire, whether blank forms for another business, broadsides or other ephemeral literature, or even pamphlets. The prices for these were calculated on a case-by-case basis. One guide is a set of notes prepared by Benjamin Franklin for his printing office in the 1750s. He advised that to determine the price of a publication, one should "Compute Journeymens' Wages at Press and Case, treble the Sum, and that is the Price per Sheet for the Work." For other types of printing, he provided itemized guidelines that suggest the range of items that printers produced:

> Blanks for Office, ½ Sheets, N° 300 and upwards, Printing 1d a Piece.
> Broadsides Ditto 2d a Piece
> Hatters Bills 25s per 1000
> Paper Money 1d per Pound, besides Paper and Cuts.

Party-Papers, Quadruple Journeymens' Wages.
Bills of Lading 6s per Quire
Apprentices Indentures 8^{d} a Pair, 6s per Day
Bonds 4^{d} single, 3s per Day. 5s per Quire
Bills of Sale 3^{d}–2s3^{d} per Day
Powers of Attorney 4d–3s per Day
Portage Bills 8^{d} each[44]

Printers would charge varying amounts for a pamphlet and a similar small publication, usually in the range of a few pennies to several shillings, with prices as high as several pounds for a book. These prices, while indicative of the range of printers' prices and fees, should also be taken as a best-case scenario; Franklin's shop was the most successful in all of North America and he clearly had a knack for business that his trade brethren did not all share.

To maintain constant visibility and (perhaps) produce steady revenue, printers typically sought to establish a newspaper for their offices. It produced two revenue streams, each related to its function as a periodical publication. One was subscriptions, for which printers typically charged eight to ten shillings per year during the late colonial period.[45] For a newspaper of average distribution (about five hundred copies per issue), that subscription could hypothetically bring in about four hundred pounds per year. But subscribers were notoriously bad at paying, and printers constantly hounded, hectored, and cajoled them into settling their accounts. William Rind, for one, spared no overwrought phrase in trying to convince his customers to pay their accounts. He claimed in the *Virginia Gazette* that "I hardly receive enough from my Gazette to pay the riders I am obliged to employ to dispense it; so that all my paper, the maintenance and pay of workmen, &c. &c, are hitherto sunk." He implored his readers to settle their accounts in order to "to save myself and family from utter ruin."[46] Advertising was by far the more lucrative part of newspaper publication, and advertisers had a somewhat better record of payment. To encourage merchants and tradesmen to advertise, printers often gave discounts for advertisements repeated over the course of several weeks, for which printers could save the set type from week to week. In the mid-1760s, for example, John Holt charged five shillings to run an advertisement for four weeks and one shilling each week thereafter, amounts that were relatively common.[47] Using that amount, the average newspaper, which included nearly two pages of advertisements each week, could have brought in several pounds of revenue per issue.

These forms of printing in combination could enable a printer to support himself and his family. But to ensure a steady income, printers looked to the government, the most reliable and consistent contractor for printing services. In each colony—and especially in its capital—various government institutions had regular printing needs. Assemblies published their proceedings and session laws. Governors and councils published proclamations and executive actions. Customs offices and courts required forms and other printed paperwork. Colonies occasionally issued paper currency to loosen the money supply. And colonial governments paid reasonably well and with regular frequency. In brief, to make money, printers tried to get the contract to print it.

Many printers therefore competed for government contracts or to earn a nod as official printer to the colony or one of its branches of government. As in the cases of Jonas Green, William Dunlap, Anne Catharine Green, and James Parker, many eagerly sought to become official printers to imperial and colonial governments. Government printing jobs provided a steady stream of income and useful political connections. These printers supported themselves by printing laws, legislative minutes, paper currency, and various proclamations and other occasional work. In smaller colonies, a town might have only one printer, and often the nongovernmental work was insufficient to sustain even this lone printing office. Each colony appointed at least one printer to print its laws, the legislature's journals, and sundry other items. At least fifty printers held an appointment with a government during the imperial crisis, either as official printers or in other public capacities (such as postmaster or secretary to a legislative body). Governments in larger colonies often divided the work, providing employment to several printers. In Pennsylvania, for example, the assembly contracted in 1769 with both Hall and Sellers and William Goddard for printing portions of their laws, paying Hall and Sellers £56, 15s., and Goddard £140, 4s., 6d. Because the colony had a significant and growing German population, the assembly also contracted in the same year for work with German printer John Henry Miller, paying him £33, 15s.[48] Several colonies split the printing work on the basis of the branch of government, with the governor and his executive council hiring one printing firm and the legislature another. These arrangements could work well for both the printers and governments involved. For the former, dividing the government work meant that they had the benefit of the steady income it provided without being overwhelmed for certain portions of the year by too large a volume of business, as could occur for example at the end of a legislative session. For the governments, it provided security in case one printer should fail, as the task could be turned over to the other printer or printers. In addition, it allowed different government factions to patronize various constituencies, as the Pennsylvania Assembly was able to do with the German-language press.

To complement their government contracts and supplement their income, printers also sought other, nonprinting positions in both the colonial and imperial administrative hierarchies. At least twenty-two printers held some sort of government position, ranging from town administrators and justices of the peace to jobs in the imperial postal system. Jonas Green rose to be clerk of the Maryland General Court, and both John and Daniel Kneeland served as assessors in Boston. Several printers were active in the post office, serving as local postmasters, which brought not only income but also first access to news from other parts of the colonies. Many received their appointments through their connection with Franklin, who served from 1753 to 1774 as deputy postmaster general for North America. Ever the assiduous cultivator of relationships, Franklin used his position to secure connections and improve his partners' enterprises. As the foremost colonist living in England in the 1760s, Franklin was also helpful in securing other posts. For James Parker he got an appointment to the customs office in 1765, as Parker claimed that an honest tradesman could not afford to live in New York City on his job alone. In fact, one imperial appointment was apparently not enough: he was already secretary of the North American post office.[49]

Just as printers sought a variety of network connections within the trade, they also sought to establish regular contact with a broad range of suppliers and clients. Paper was a vital resource for printers that, while not quite as expensive as the press or types, still required commercial skill and good connections to obtain. Until the middle of the nineteenth century, paper was manufactured in a mill from rags and was almost always in short supply. Printers therefore remained closely tied to papermakers and often took advantage of opportunities to get involved in paper mills themselves.[50] Other printers continued to buy high-quality paper from England. David Hall had just placed a major order with his English agent when the Stamp Act was enacted, leaving him with what he thought would be uselessly unstamped paper.[51] Either way, printers acted in close coordination with their business associates. When printers themselves elected not to sell stationery supplies in their own shops, they sought out stationers as repeat customers, selling them a variety of printed goods, including blank forms for everything from indentures to legal contracts and bills of lading for ships entering the harbor.

Engravers too were vital for a printer's business, supplying woodcuts for advertisements, almanacs, newspaper mastheads, and other publications.[52]

The printing trade offered a precarious existence to its participants; ventures failed commercially as often as they succeeded, if not more so. When economic difficulties arose, printers interpreted them within a highly moralized eighteenth-century framework of business standards that emphasized personal virtue. If their commercial ventures soured, they would find fault with other people involved in their operations rather than with impersonal market forces.[53] Such was the case in Philadelphia when the eighteen-year partnership between David Hall and Benjamin Franklin ended. The political allies of Franklin in the Pennsylvania Assembly had become dissatisfied with Hall, their official printer. Hall's insistence on maintaining political neutrality at a time when the legislature and the colony's proprietors were at loggerheads made him a poor choice.[54] Joseph Galloway, speaker of the assembly, and Thomas Wharton, a prominent Philadelphia merchant, made arrangements through Franklin's associate James Parker to hire William Goddard, then working as a journeyman in Parker and Holt's New York office, as a master printer for a new, more politically compliant printing office they planned to set up. Galloway and Wharton promised Goddard a half stake in the partnership, with each of them taking a quarter. The contract also contained an option for Franklin himself to buy a stake.[55] The partnership gave the assembly a new official printer and allowed Goddard to begin publishing the *Pennsylvania Chronicle* in January 1767. Within a month of issuing the first number, he claimed more than one thousand subscribers, making it almost immediately one of the most widely subscribed newspapers in North America.[56]

Galloway and Wharton's high hopes for the new printing office were soon dashed by Goddard's independent streak. Goddard was among the more combustible men in the printing trade and perhaps the most indefatigable. Between 1762, when he opened his first printing office, and the end of the American Revolution in 1783, he operated printing offices in Providence, Philadelphia, and Baltimore, worked as a journeyman in New York and Woodbridge, New Jersey, attempted to create a new postal system in the colonies, worked for the first Continental post office (with the last two jobs requiring him to travel the full length of the Atlantic seaboard), and was nearly beaten to death by a mob in Baltimore in 1777. Goddard's attempt to run the office independently and, in particular, to control the editorial content of the *Pennsylvania Chronicle*, quickly provoked conflict among the partners. His assertions of independence incensed Galloway and Wharton, who had intended the paper as a counterweight to the others in the city (Hall and Sellers's *Pennsylvania Gazette* and Bradford's *Pennsylvania Journal*), which they considered insufficiently antiproprietary.

The partners battled over the day-to-day management of the office's cash flow. Goddard claimed that Galloway and Wharton, who were to put up much of the capital for the office, provided him with just enough funding to operate the office but never quite enough to allow him to run the office to its fullest potential. Desperate for money, he appealed to others for cash, including John Smith, a merchant in Burlington, New Jersey, from whom he sought a loan of one hundred pounds (Smith declined), and he took out a bond for sixty pounds from Franklin.[57] The silent partners then forced Goddard to take on a printing partner, Benjamin Towne. Goddard immediately suspected that Towne was a spy for Galloway and Wharton and tried to cut him out of as much business as possible. By 1770, matters had grown so unbearable to Goddard that he took the unusual step of publishing two pamphlets to denounce his partners. He titled his work, *The Partnership: Or the History of the Rise and Progress of the Pennsylvania Chronicle, &c. Wherein the Conduct of* JOSEPH GALLOWAY, *Esq; Speaker of the Honourable House of Representatives of the Province of* Pennsylvania, *Mr.* THOMAS WHARTON, *sen. and* their Man BENJAMIN TOWNE, *my late Partners with my own, is properly delineated, and their Calumnies against me fully refuted*. In the seventy-two-page pamphlet, Goddard lashed out at some of the most powerful men in Pennsylvania under his own name, laying out his grievances in great detail, printing their partnership agreement and several letters he exchanged with them and with others about the affair (including his mother), and recounting several conversations, all intended to embarrass them.[58]

Goddard's arrival in Philadelphia also caused problems for the remnants of Franklin's network, most notably David Hall. A cautious man but not a dupe, Hall immediately suspected when Goddard arrived that either Franklin or his friend William Strahan had been involved somehow and felt betrayed. He was correct on both counts, though apparently he never confirmed the truth. Franklin and Strahan both denied it, but Franklin was instrumental in bringing Goddard to Philadelphia, and Strahan purchased and shipped a set of types for Goddard's new printing office—a debt he was still trying to collect (from Galloway, not Goddard) as late as 1772.[59] Hall, meanwhile, tried to manage as best he could. In letter after letter to Strahan, he begged his London friend to send him political news for the *Gazette* with as much dispatch as possible.[60] Knowing that money was flowing elsewhere, Hall relied on decades of friendship and collegiality to ensure that he could at least continue to hold on to his information advantage over his new rival.

Printing was a constant struggle to survive. To succeed, printers needed to present themselves as men of good credit, on whom men of means could depend to pay back their debts. In turn, they had to rely on a wide range of economic actors to supply goods and pay debts to them—purveyors of the raw materials of printing on the one hand and customers on the other. Just the act of opening an office put printers deeply in debt, which would take years to repay. The economics of printing, therefore, was not particularly amenable to the men and women of the trade, who had to navigate precarious business conditions as well as difficult social and political situations.

Printers and Atlantic Information Networks

Financial capital was not the only resource of which printers required a constant supply. They also needed a steady stream of information: texts that authors wanted to publish, news from far-flung locales to include in their newspapers, geographic and meteorological calculations for their almanacs. To ensure that access, printers needed contacts, preferably well connected themselves, who could relay news and information to them, both within the town where they worked and across the Atlantic Ocean. Like merchants and other businessmen with far-flung interests, therefore, printers needed to establish a range of connections around the Atlantic world to increase their access to the most current information—their most precious commodity. These networks were extraordinarily common in the eighteenth-century Atlantic world. They encompassed economic endeavors, such as the slave trade and trade in commodities such as wine, sugar, tobacco, and chocolate, as well as social organizations, scientific projects, and other cultural efforts.[61] These connections were not closed circuits among a relatively fixed set of commercial actors, as they would become in the modern era, but were highly personal, allowing each actor to collaborate with a wide variety of networks in a fluid and ever-changing environment.[62] Printers sought to establish connections with three categories of people: other members of the printing trade, local elites, and long-distance information suppliers.

Benjamin Franklin, who himself was introduced to the trade through family as an apprentice for his brother James in Boston, had no equal in establishing affiliated and satellite printing offices throughout the colonies.[63] He made a practice of establishing partnerships in other locations with promising former apprentices and journeymen (map 2). In doing so, he ensured himself the ability to make money in new colonial markets and moved potential competition to other places. Franklin made his first agreement less than five years after he had established himself in Philadelphia as a master printer, sending his onetime journeyman Louis Timothée (later Lewis Timothy) to establish a printing office in Charleston, South Carolina, in response to a request from the colony's General Assembly for a printer to the colony. After Timothy's death, that agreement continued with both his widow Elizabeth and then their son, Peter. During his long career, Franklin would also set up printing offices in New York, New Haven, Antigua, Dominica, Jamaica, and Barbados. In addition, he set up several partnerships in Pennsylvania with German printers in an attempt to make his enterprises multilingual.[64] Even though Franklin retired as an active printer in 1748, and his financial stake in his own printing office ended in 1766 when he dissolved his partnership with David Hall, he maintained throughout his life an enormous influence over many of the members of his network. In fact, he corresponded with his colleagues continuously about matters related to their printing businesses and their associations with the post office in the last decades of his life.

Before the Revolution, few replicated Franklin's entrepreneurial model. Much more common was a network in which printers joined together in either ad hoc or long-term relationships for their mutual commercial benefit. These connections could prove crucial to the success or failure of a venture. Printers also sought to pool their resources and minimize risk whenever possible for individual projects. Such combinations occurred most frequently when a project required a greater investment of capital than any single office could muster. For example, printers often cooperated on book projects because they would have overwhelmed the labor and materials of any one office. Co-publication also spread the risk of investing large sums of capital across multiple printing offices. In 1773 John Holt posted an advertisement in his *New-York Journal* for a collection of church music about to be published. He utilized his network of Franklin contacts to gather subscriptions, listing such printers as John Dunlap of Philadelphia and Alexander Purdie and John Dixon of Williamsburg and his connections to the Green family through Richard Draper of Boston and Frederick Green of Annapolis. He also agreed to collaborate with his rival printers in New York, Hugh Gaine and James Rivington.[65] Printers also frequently shared publication costs with other members of the book trades, in particular booksellers.[66]

Collaboration could take many forms. For instance, a group of printers in a single town would often band together to co-publish an almanac, a steady

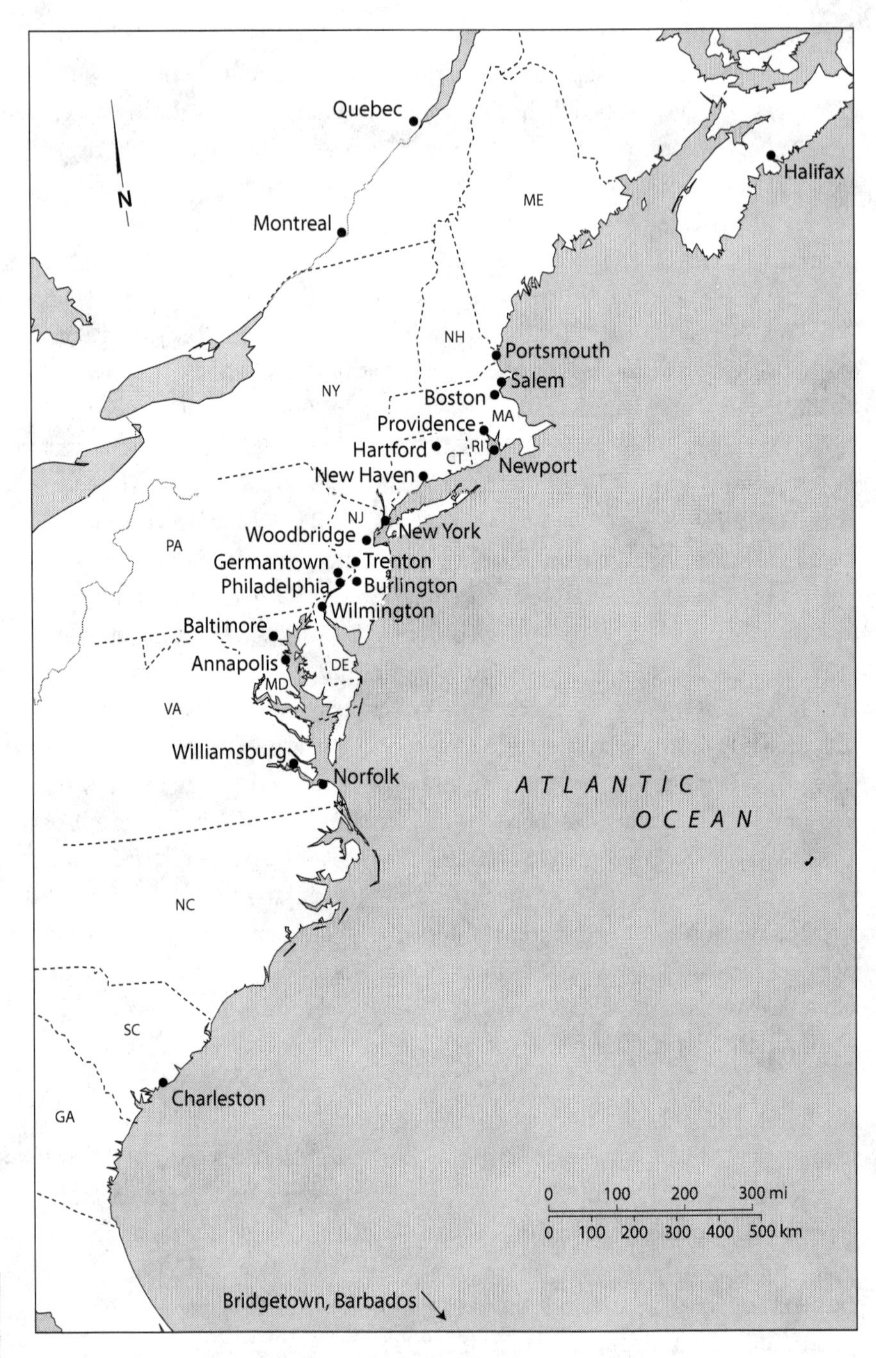

Map 2. Active printers with a relationship with Benjamin Franklin, ca. 1763–76

seller that made money for everyone. *Ames's Almanac* is probably the best example of that phenomenon. Throughout the 1760s and 1770s, nearly all the printing offices in Boston cooperated in publishing this almanac.[67] Such collaboration saved the printers the trouble of competing against one another so that everyone made money and circumvented the habitual practice of pirating almanacs, which caused numerous disputes in the precopyright era.[68] Printing was therefore not seen as a zero-sum game by its practitioners. In fact, printers would open the pages of their publications, in particular their newspapers, so that other printers could advertise the establishment of a new newspaper in their pages.

A master's contacts could sometimes clear the path for a printer to start his own office, but young printers often developed ties to multiple networks. These connections were created by a trade system designed to move young printers through a series of jobs before they became established in business on their own or in partnership with others. A printer often apprenticed for one master and then perhaps worked as a journeyman for one or more other printers before finally settling into his own office. At each step, he could work with people who had connections to other networks. If a printer achieved the status of a master, he could then train his own apprentices and hire journeymen, renewing the cycle. In so doing, each printer would make family and other connections that could prove useful in his career.

Creating personal connections from scratch was disadvantageous, and printers were loathe to do it. For example, Isaiah Thomas in 1769 applied to work with Adam Boyd, who had set up a printing office in Wilmington, North Carolina. Boyd replied that he would happily partner with Thomas but "could not give [him] the Encouragement [he] want[ed]." Furthermore, Boyd wrote, he would not want to take on a partner: "The Times are very critical & at all Times the Director of a printing office is liable to Censure & when this would happen You would like as little to hear Censure for Me as I would for you." In other words, the tense political situation might quickly reveal ideological differences between the would-be partners that could prove difficult to reconcile. While he thought Thomas was "properly qualified for the Business," the situation might be different if they were not "Strangers to each other."[69] Thomas also dissolved his partnership with Zechariah Fowle within two years, arguing that printing would "not maintain two families and pay such heavy debts." He did, however, agree to "employ your two Boys and learn them the Art, if we can agree to that matter."[70] Thomas's low opinion of Fowle's skill as a printer likely also hampered their joint prospects.

Printers looked beyond the printing trade to extend their connections by seeking to integrate themselves in the associational life of their communities. For example, they eagerly waded into the religious and cultural life of their towns. In New York, Hugh Gaine made regular payments to rent a pew at St. Paul's Chapel and in 1769 made a payment of £1, 10s. "for three years Subscription to the New York society Library Due in May Last."[71] While the religious inclinations of most printers are unknown, they frequently published religious tracts, sermons, and essays and rarely shied away from controversies that could produce more printed debate. Religious topics dominated the non-newspaper imprints of early America and frequently proved lucrative for printers. Printers also took part in the secular life of their communities, joining such voluntary associations as subscription libraries, fire companies, and the Freemasons. In Philadelphia, printers joined a diverse group of associations, including library companies, fire companies, the Colony in Schuylkill, and ethnic groups such as the St. Andrew's Society and the German Society.[72] Franklin, who founded a variety of associations in Philadelphia, is not the exception but the archetype of the community-minded printer. Peter Timothy of Charleston was a Freemason and a founder of that city's Library Society. Printers sought to be a part of their communities in many ways, whether for their own enrichment, to cultivate connections, or both.

The third group that printers sought to connect with can be colloquially termed "information nodes." These people had access to information—the "freshest advices" that every printer promised would appear in his newspaper—and transported it across the Atlantic and along the North American coastline. Because eighteenth-century newspapers relied more heavily on imperial than local news, printers worked hard to develop a broad array of connections both in North America and around the Atlantic. Close contact with ship captains and common mariners created a steady supply of news from ports around the Atlantic rim, including not only England but Continental Europe and the West Indies. Formal connections between correspondents dedicated to the transmission of information, like that between David Hall and William Strahan, were not common but provided interesting news faster than through the standard method of reprinting items from the English newspapers, which almost all printers received. Printers also received news from third parties who had Atlantic contacts, like merchants who would deliver the newsworthy or public parts of their correspondents' letters to the printing office for publication. As for North American news, printers attempted to maintain good relations with their local postmasters in order to protect their access to other colonial newspapers. Many printers themselves —sixty-seven of those in the study—sought appointments as postmasters for this very reason, as well as to enhance their ability to distribute their own products. Finally, printers needed to be on good terms with authors and commercial men within their communities, not to mention staying up to date on the latest debates in local taverns and coffeehouses to ensure a supply of items and advertisements for their newspapers.

In addition to individual connections, printers attempted to link into institutional pathways through which they could facilitate rapid communication. They devoted a great deal of energy to the post office, even though it was founded with different interests at its core. In its design, the imperial post office served three government objectives.[73] First, Parliament explicitly intended the post office to generate revenue for the Exchequer, to the unrealistic tune of seven hundred pounds per week beginning in 1711.[74] Second, Parliament designed the imperial postal system (and the General Post Office in London that operated it) to facilitate imperial communication, that is, the correspondence that made the administration of empire possible. In that regard, the imperial postal system was largely indifferent to the communication and information needs of printers along the North American coastline, preferring instead to emphasize communication between London as the hub and various towns as the spokes. That neglect began to shift somewhat in the 1760s, however, as Parliament reorganized the post office as part of its efforts to streamline imperial administration after the end of the Seven Years' War.[75] Third, the Post Office sought to assert control over the flow of politically sensitive information and to provide surveillance of groups and individuals who opposed the government. This practice came to the fore during the reign of King Charles II. Before the 1760s, there were few documented cases of intercepted correspondence in or from the colonies, but colonists were nonetheless attuned to the possibility of an assault on what they saw as private property.[76] Despite these drawbacks, over the course of the eighteenth century the post office grew from a rickety provincial scheme with just a few riders into an important institution within the British imperial communications network.

The post office did not serve all of printers' long-distance information needs, however, because its structure and design limited its effectiveness as a tool of communication. Most importantly, the post office connected primarily the coastal towns in the colonies. Further inland, communication relied on the British Army, Native Americans, traders, and other personal messengers.[77] In addition, mailing a letter was quite expensive. Before 1765, a letter cost one shilling per sheet of paper between New York and Boston, and two and a half shillings from Boston to Charleston, more than a week's wages for an ordinary laborer. Even after Parliament reduced postage rates in 1765, the cost of sending a letter remained extremely high for most people.[78] Thus, use of the post office was generally limited to printers, who needed it to circulate their products, and merchants, who required a reliable means to correspond about their businesses and receive information about shipping and commerce.[79] Other colonists exploited loopholes in postal laws and regulations to avoid the legal monopoly and high postage rates. Many colonists simply did not use the post, preferring to send letters via traveling friends or servants. In Newport, for instance, surveyor Hugh Finlay found two parallel postal systems. In a conversation with Rhode Island postmaster Thomas Vernon, Finlay learned that there were "two Post offices in New Port, the King's and Mumfords." Benjamin Mumford was the imperial post rider who, according to Finlay, carried a great deal of mail on his personal account.[80] Finlay noted that the post office lost a significant amount of business to stagecoaches that ran routes along New England post roads and discovered that "it is the constant practice of all the riders between New York and Boston to defraud the Revenue as much as they can in pocketing the postage of all way letters." He complained that riders were diverting so much business that they had made "pack beasts of the horses which carry His Majesty's Mails."[81] Rates for mailing parcels were low compared to letters, and colonists took advantage by attaching letters to packages that were sometimes no more than "little bundles of chips, straw, or old paper."[82] In utilizing alternative pathways, colonists created webs of connection that were more comprehensive and had greater reach than the imperial post office.

Nonetheless, printers were among the post office's primary users, and their commercial print operations relied upon the postal system to gather and circulate news and newspapers. The post office was a vital link in the interdependent world of printers and the news they circulated.[83] Its infrastructure was crucial to the intercolonial commercial networks that printers had created in the North American colonies, both on the mainland and in the West Indies. The production of newspapers relied on the circulation of news among printers for fresh content and sales depended on the timely distribution of the newspapers themselves. And, of course, printers circulated their newspapers as a for-profit venture.

Printers therefore used the post office for their own commercial purposes and sought financial advantage by participating as postmasters in its administration. A position as local postmaster ensured a printer of first access not only to news coming from other colonial towns but often to the earliest warning of news from England, a key component of eighteenth-century newspapers. Finally, in the early part of the century, the position came with the franking privilege, which allowed the postmaster to send and receive mail without paying postage. This perk was vital for printers who relied so heavily on newspaper exchanges among themselves as sources of news.[84] The onset of winter, for example, had an enormous impact on the content of newspapers: reports from distant northern towns were often replaced with notations that post riders had not reached the printer's office before the newspaper was printed. Just after Christmas in 1773, for example, James Rivington of New York noted that "all the several posts are become irregular in their arrival and departure, owing to the tempestuous weather last week, and to the ample and very seasonable fall of snow, with a welcome succession of frost, that we have had this week."[85] Even as other colonists bypassed official channels of communication, printers kept careful watch on the post office and the opportunities it presented them.

The "Free and Open Press" as Ideology and Business Strategy

Key to printers' projection of their place in business and politics was their collective ideology of the press. Colonial and Revolutionary printers, in fact, frequently fostered the perception that they did not matter (a notion that historians have extended by taking it at face value). Many printers described themselves as "meer mechanics," that is, as artisans who set type and pulled the press but nothing more. In large part that strategy reigned during the colonial period as a fairly logical business reaction to a regulatory regime that protected powerful interests with a broad definition of and harsh penalties for sedition and libel. For the most part, colonial American printers before the imperial crisis relied heavily on the habits of English printers as a matter of precedent, though their businesses differed from those of their English counterparts in key respects (not least their size).

Printers in both England and America had long advocated for a doctrine of a "free and open press," which meant that the press would be both "free" of government restrictions and "open" as a forum for debate in the public interest. In England, regulation of the press had long been contentious as both Parliament and successive monarchs sought to circumscribe the power of the press and printers. The most important method was through the Stationers'

Company and a series of Licensing Acts. Until 1695, the acts limited printing at least in theory to members of the Stationers' Company, a tightly controlled trade guild, who could operate presses only in London and the university towns of Oxford and Cambridge. These attempts at state regulation gradually became less effective during the course of the seventeenth century as the printing trade grew. Direct challenges also arose. John Milton, for example, protested the licensing of publications in *Areopagitica*, a 1644 publication addressed to the Long Parliament.[86] The diminution of restrictions on publication culminated with the lapsing of the Licensing Act in 1695. For all intents and purposes, the demise of the law ended in England the practice of prior restraint of printing.

Colonists buttressed their claims to a free press through a theory that scholars have described as opposition ideology. Eighteenth-century British North American colonists absorbed a great deal of political theory from seventeenth-century republicans, some of whose names became common pseudonyms in colonial essays: James Harrington, John Hampden, and Algernon Sidney, along with the early eighteenth-century "Cato." These writers suggested that the political world functioned as a constant struggle for power. In order to hold and consolidate power, the theory held, officials were always at risk of becoming corrupt and attempting to curtail people's liberty. The people, they argued, must defend their liberty through various means and advocated a government structure that limited its power. One key facet of that effort was a free press—one that could expose the corruption and wrongdoing of the government. The news from Britain that colonists read and discussed during the first half of the eighteenth century seemed to confirm their suspicions that government was corrupt and that they must always be on guard. During the ministry of Robert Walpole (1721–42), these writers argued that the government remained stable only through the corruption of patronage: the government silenced leading critics simply by co-opting them. The South Sea Company scandal in the early 1720s revealed staggering corruption. The scandal prompted James Trenchard and Thomas Gordon to publish a series of essays under the pseudonym "Cato."[87] The letters—widely reprinted in North America during the imperial crisis several decades later—argued that the press operated as a "bulwark against governmental power."[88] Given their structural disadvantage within the political system, the people required recourse to the press in order to expose the corrupt actions of the government whenever it threatened the people's liberty.[89]

Printers asserted the principle in the defense of the free press but could not stop governments from attempting to shape political communications. The standard account of controversies in North America begins in 1690, when Benjamin Harris published the first newspaper in the English colonies, *Publick Occurrences*, in Boston. Authorities in Massachusetts suppressed the newspaper after only one issue because he had reported on political events within the colony and spread gossip about King Louis XIV of France. In the 1720s, Benjamin Franklin and his older brother James got into trouble during a controversy over smallpox inoculation. James Franklin, publisher of the *New England Courant*, made the paper a forum for criticizing the pro-inoculation strategy urged by local civil and religious authorities, including Reverend Cotton Mather. Because he was not publishing "by authority," the colonial government imprisoned James, ostensibly for criticizing the government's approach to piracy.[90] He was released after being warned not to publish on the issue and soon left Boston for Newport, Rhode Island.

Boston broke the ground, but New York produced the case most famous for its defense of a free press, even though scholars now agree that its impact as a legal precedent was minimal. The tale of John Peter Zenger, a German immigrant who had come to New York to train as a printer, began as a political story among warring factions in New York colonial politics. A new governor, William Cosby, arrived in 1732 and soon entered a dispute with the chief justice, Lewis Morris, whom he discharged. Morris, among the richest men in the colony, led a group of disaffected Dutch and English colonists against Cosby, aiming to drive him from New York. Because William Bradford, who published the only newspaper in New York at the time (the *New-York Gazette*), printed under government authority, the Morris faction started a newspaper of its own in 1733 (the *New-York Weekly Journal*) and hired Zenger to print it.[91] The newspaper included a variety of political essays, in particular the writings of British authors Addison, Steele, Trenchard, and Gordon on the dangers of corruption. Zenger also occasionally published "sham advertisements which abused the Cosbyites in the grossest terms," comparing the governor to a monkey and his advisers to lost puppies.[92]

Within a few months, Bradford and the *Gazette* engaged with the arguments in the *Weekly Journal* in defense of the governor, sparking a newspaper war. The two publications debated the virtues of freedom of the press as weighed against the law of libel at a time when Cosby sensed that the opposition was gaining in popularity. Through the newly promoted chief justice, James De Lancey, Cosby attempted to prosecute Zenger on charges of libel, but grand juries composed of local residents refused to comply. The governor's council, however, did oblige, ordering several issues of the *Weekly Journal* burned for seditious libel and Zenger imprisoned for printing them. Zenger's initial defense attorneys, part of the Morrisite faction, were disbarred by De Lancey, so his supporters hired Andrew Hamilton of Philadelphia to defend him. Making an argument that defied existing statute, Hamilton contended that the key to libel was its veracity rather than its intended target—in other words, that speaking the truth was an adequate defense against a libel charge. The jury agreed and found Zenger not guilty.[93] The decision, however, did not establish a legal precedent, nor did it protect printers in the following decades from prosecution. Yet it remains the most famous case involving the freedom of the press in colonial America.

To counter the political pressures on their businesses, colonial American printers developed a business strategy based on the principle of the "free and open press."[94] Benjamin Franklin gave the clearest exposition of the free press doctrine as it existed in the eighteenth century in response to an attack against a broadside advertisement he had published in 1731 (the ad itself has not survived). Franklin's response to the criticism for publishing the advertisement, which had been posted throughout Philadelphia, was to proclaim his neutrality with respect to the content of his publications. In a long address to his readers, which has since become known as his "Apology for Printers," Franklin attempted to explain his position. He contended first that "the Business of Printing has chiefly to do with Mens Opinions; most things that are printed tending to promote some, or oppose others." Therefore, there was no way for a printer to avoid conflict entirely. Any opinion offered might be opposed, and sometimes strenuously. For Franklin, the solution was for printers to be disinterested. They should "acquire a vast Unconcernedness as to the right or wrong Opinions contain'd in what they print" and should print the views of all comers, not just those with which they agreed. Likewise, readers should not hold printers accountable for the views they printed. Furthermore, he argued that printers already censored on their own a variety of material they considered unsuitable. Franklin noted—perhaps with an air of superiority, and certainly with an eye toward his customers—that he had "constantly refused to print any thing that might countenance Vice, or promote Immorality," although he might have earned considerable profit by doing so in view of "the corrupt Taste of the Majority."[95]

This line of reasoning tied the printers' commercial interests directly to their ability to appear as impartial observers to the debates occurring within the pages of their own publications. The rhetoric of the free press suggested that printers were merely providing all sides to the story or merely responding to public demands, and printers adhered to the doctrine for much of the eighteenth century in the hopes of immunizing themselves against the wrath of political incumbents. Even during the imperial crisis, emergent newspapers and printers defined themselves in terms of serving the public good rather than a partisan cause.[96] Samuel Inslee and Anthony Carr took over James Parker's printing office in New York at his death in 1770, continued his *New-York Gazette: or, the Weekly Post-Boy*, and used the doctrine of the free press to excuse their being "*young Beginners*." They promised that the paper would "be sacred to the Cause of Truth and Liberty, and never be prostituted to the Purposes of Party, but be equally free for all . . . to have their Productions inserted."[97] The printers of the *Quebec Gazette* made similar promises in 1764, asking that "*this one thing we beg may be believed, That* Party Prejudice, *or* private Scandal, *will never find a Place in this* Paper."[98] For most printers, appeals of this sort were commonplace.

Printers put forward the idea of a free press as a commercial strategy to place themselves outside the realm of political combat, but their decision had enormous political impact. In doing so, they disavowed ultimate authority over the political content of their newspapers and attempted to shift responsibility to the authors of pieces, who were usually anonymous, or to the field of public opinion. For many, this action was deliberate and necessary to protect a fragile printing market. After 1765 this strategy became increasingly untenable and unattractive for printers as the imperial crisis shifted the landscape of commercial interests for them.

Colonial printers were linked to one another through trade networks that funneled younger printers through their apprenticeships into employment in a variety of printing offices and, if fortunate, toward the proprietorship of offices on their own. Their success depended not only on their skills but also on their ability to develop effective credit mechanisms to open and sustain their operations. To do that, printers cultivated contacts within the printing trade, setting up partnerships between masters and former apprentices and creating kinship ties through marriage. In this they closely resembled other commercial men and women. Likewise, they developed contacts in allied trades such as papermaking and engraving and sought to serve people and institutions that generated steady demand for printing work, especially local, colonial, and imperial governments. They also sought out members of the British Atlantic elite as suppliers of information, the most important raw material of publishing.

For many printers, success was either unattainable or fleeting. But for a core group of longtime participants in the trade, access to a broad range of contacts made them central actors in the commercial and political life of their communities, even as they usually portrayed themselves as politically neutral. They were among the members of their communities who had the densest and most diverse social ties, connections ranging from sailors and poor apprentices who might supply tidbits of information to the leading merchants of the town, who were sources of both information and customers. Their contacts stretched beyond the localities in which they worked and could extend as far as London. Printers were able to manage these diverse contacts in part because of their unique socioeconomic position within their communities. On the one hand, they were skilled artisans who produced a physical object. Such work put them clearly among the community of other skilled artisans in urban settings, some of whom had considerable financial means but most of whom were very much middling. At the same time, printers had access to the upper echelons of colonial society. Through their solicitation of advertisements and publication of belles lettres, political tracts, and other literary endeavors, printers conversed regularly with the commercial and political elite. They sought patronage appointments through the local, colonial, and imperial governments. And they had to be well educated and culturally literate in order to participate in the transatlantic conversation occurring in print.

The printers who managed these varying challenges successfully were those who could best adapt to changing circumstances and retain contacts with politically relevant social groups. Colonial towns and cities, even as they grew substantially during the eighteenth century, remained for the most part small and close-knit places. Because of the emphasis on personal credit and reputation, the connections that printers cultivated frequently proved vital to a printer's commercial survival. A broad base of support that cut across factions within a community was helpful not only for a printer's access to information but also as a buttress against controversy and difficult financial times. Therefore, printers portrayed themselves publicly as agents of the public, providing a service to the community. They were not, they argued, interested parties to political activities and disputes aired in the pages of their publications. Instead, they were impartial or disinterested observers who facilitated debate, enforcing only those rules and guidelines that maintained order and decorum. In other words, they sought to make themselves the indispensable arbiters of public debate in colonial America.

A Trade under Threat

Printers and the Stamp Act Crisis

In February 1765, just a week after the Stamp Act had passed the House of Commons, Benjamin Franklin, the Pennsylvania Assembly's agent in London, penned several letters to inform his compatriots in America of the unfortunate news. He recognized immediately that the new law would have a profound impact on the printing trade. "Every Newspaper, Advertisement and Almanack is severely tax'd," he wrote to his friend John Ross. "If this should, as I imagine it will, occasion less Law, and less Printing, 'twill fall particularly hard on us Lawyers and Printers."[1] Indeed, the act proposed taxes on nearly every kind of document produced in a printing office, including not only preprinted business and legal forms but also its mainstays, newspapers and almanacs and the advertisements that appeared in them. Printers found their businesses at the center of protests that stalled economic life in America for nearly a year in what would become the first concerted resistance to imperial policies that led to Revolution. They produced the pamphlets written by prominent men such as Daniel Dulany and James Otis, composed and edited chronicles of riots in their towns, and selected and reprinted accounts and essays from newspapers in other cities. Printers and their businesses also became the objects of protest, as their decisions whether to continue to print, with or without stamps, became a politically charged issue.

Like the Revolution more generally, the Stamp Act has not lacked for chroniclers and interpreters. Historians have examined the Stamp Act crisis from a variety of perspectives but almost always with a focus on the political ramifications of the new British imperial policy. Scholars of revolutionary ideology have vigorously examined the debates that emerged about the validity of parliamentary taxation and the broader question of Parliament's right to legislate for the colonies.[2] These arguments, per the standard narrative, promulgated throughout the colonies in pamphlets and newspapers and convinced many colonists of the need to resist the Stamp Act. Accordingly, colonists effectively nullified the law through massive street actions in thirteen of Britain's twenty-six North American colonies. The violence was, according to one historian, "the most ferocious attack on private property in the history of the English colonies," lasting throughout much of the second half of 1765 and first few months of 1766.[3] These scholars have nodded toward colonial print media and their influence on the protests. Yet print remains in their narratives a static background to the larger story of resistance. In fact, the best-known chroniclers of the Stamp Act crisis, Edmund and Helen Morgan, hardly mention either printers or their publications in their foundational work on the subject.[4]

Those scholars who have pointed to 1765 have argued that printers took political positions only at the behest of others. These accounts, which rely heavily on the writings of Franklin's associates and a few others, suggest that printers could either take the Patriot stance (often under direct threat from the crowd) or become de facto loyalists should they choose to maintain some semblance of neutrality. In other words, printers were buffeted by forces beyond their control into taking a side that they may have supported only lukewarmly.[5] Using a more comprehensive review of the members of the printing trade, this chapter instead argues that printers displayed a much broader range of reactions to the Stamp Act, from indignant loyalism to enthusiastic militance. Some printers, to be sure, were forced to take sides, but many actively worked to advance their own commercial interests and political viewpoints and to shape public opinion. In so doing, they sought to balance their commercial and political self-interest, their commitment to the idea of a free press, and their sense of printing as a public service.

The Stamp Act crisis can be understood fully only by placing the political debates that proliferated in print, and even in the streets, within the context of printing as a business enterprise. Because printers lacked foreknowledge of whether British authorities would be able to enforce the law, their decisions about how they addressed the political questions of 1765 were fraught with danger for their businesses. It is therefore all the more compelling to examine printers during the Stamp Act crisis. They controlled the means of communication, crafting the printed texts that disseminated information about protests, riots, and the forced resignations of stamp officers. They brought to bear in that effort their technical expertise in producing visual and printed propaganda and made their newspapers and other printed documents a public forum for debating the Stamp Act.

The Stamp Act and the Printing Trade

The Stamp Act was the brainchild of George Grenville, chancellor of the exchequer from 1763 to 1765. At the end of the Seven Years' War, Great Britain faced massive debt, the literal price to be paid for becoming the dominant imperial power in North America. William Pitt, the "Great Commoner," had led Britain to victory, and Grenville was charged with paying the debt that Pitt had amassed. Making a calculation on the basis of his metropolitan perspective, Grenville reasoned that the North American colonists, who paid the lowest taxes in the empire, should help eliminate the debt, especially because Parliament planned new expenditures—most importantly for the defense of the North American frontier—that would directly benefit the colonists. In addition, Parliament after 1763 began a broad project of imperial reorganization, streamlining the administration of empire in terms of customs, the military, the post office, and other arenas of government, as well as revisiting decades-old trade regulations.[6] As part of the restructuring, Parliament aimed to reform both taxation and the management of the colonial economy. To address these issues, Parliament in 1763 and 1764 passed two acts directed at the colonies. The first, the Currency Act, forbade colonies from issuing their own paper money, with an exception for New England. Many colonial governments had issued paper currency as legal tender as a way to maintain cash flow in specie-poor environments, in particular during the Seven Years' War. London merchants, on the other hand, believed colonists were defrauding them by artificially inflating their colonial paper to escape their debts.[7] The second measure was the American Duties Act, more commonly known as the Sugar Act. This law included several provisions to improve customs collection and to force colonists away from rum distilled in French colonies.[8] Though the Sugar Act in particular would later join the litany of tax acts that Americans consider parliamentary overreach, response at the time was muted, both from the colonial government and in print. Both acts prompted some discussion in newspapers but no large-scale protests. Those essayists who took up their pens in protest found mostly local audiences for their arguments.[9] Boston radical leader James Otis, for example, published a now-famous pamphlet with Edes and Gill that claimed the taxation was unconstitutional. All four Boston newspapers advertised the pamphlet, but it appeared nowhere else in the colonies and prompted no discussion.[10]

At the same time, Parliament began work on a stamp duty that would require most legal and business documents and most printed matter to bear a stamp purchased from the government. In letters to the governors of each colony, Grenville solicited both comments on which documents could most readily sustain a duty and suggestions for other ways to raise funds. It was an offer he knew they would refuse. While no colonies developed a plan for self-taxation, four did send petitions to the Crown objecting to the duty.[11] Colonial agents, including Benjamin Franklin, objected to the plan as well, but in February 1765 Parliament passed the Stamp Act. It received the assent of George III shortly thereafter.

The act was remarkably comprehensive in its reach, affecting nearly all aspects of public life. With duties ranging from a few pennies to several pounds per item, the act placed a duty on all printed documents, including legal proceedings, indentures, business documents, petitions, bills of lading, and even items such as cards and dice.[12] Newspapers, pamphlets, and almanacs were taxed according to their size. To collect these duties, Parliament ordered the appointment of stamp commissioners for each colony. The Stamp Act was particularly menacing for printers, as it imposed significant costs on almost every aspect of their businesses and did so in advance of the sale of their goods. For nearly all items, printers were required to pay the duty by purchasing prestamped paper from the commissioners—and to pay in sterling rather than colonial currency. For newspapers, the act also imposed a two-shilling tax on every advertisement that appeared in a newspaper.[13] These ads were a great source of revenue for colonial printers. All told, Parliament initially believed it might raise about £100,000 each year, with the possibility of more as Americans grew accustomed to the measure.[14]

A similar stamp act had been in effect in Britain since 1712, which produced annual revenue in excess of forty-three thousand pounds from newspapers and three thousand pounds from almanacs by the late 1750s. In London and elsewhere in Britain, however, printers often adjusted their practices to minimize their liability—for example, by publishing their newspapers as "pamphlets," which qualified under a different (and more lightly taxed) category of the law. When that loophole was closed, they reduced their editions to four pages, below the threshold for the duty. The British stamp acts also led London printers to experiment with smaller typefaces to fit more print on the page and with multiple columns of type on each page. Even so, by 1775

British printers were purchasing more than twelve million newspaper stamps per year.[15] But their newspapers carried a high price tag, and subscriptions were still largely confined to elites.

American colonists were therefore familiar with the idea of a stamp duty. In addition to the tax in England, colonial assemblies had occasionally used stamps to generate revenue. During the Seven Years' War, both New York and Massachusetts had passed temporary stamp acts to raise money to muster troops in defense of their frontiers, although some printers, notably James Parker, had opposed these measures. Jamaica also had passed a local stamp act to support a defense force in the wake of a 1760 slave rebellion on the island.[16] The New York law appointed two colonists—Abraham Lott Jr. and Isaac Low, both merchants, who would later oppose the 1765 act—to supervise a stamp office that would distribute stamps ranging in value from a halfpenny to four pence. Its range was not nearly so comprehensive as the 1765 Act of Parliament, but the duty did apply to such documents as newspapers, indentures, deeds and mortgages, bills of lading, and various court papers.[17] In 1765 colonists, including printers, would minimize the political significance of their acquiescence to these earlier laws by making what became the canonical argument that citizens could be taxed only by their duly elected representatives: the Massachusetts and New York legislatures could pass a stamp act for the colonies, but Parliament could not.

North American printers were in a precarious position in 1765 because the additional stamp duty on their products threatened to destroy the fragile balance they had constructed to survive. Unlike their brethren in many other craft trades and unlike most laborers, they faced grave and immediate danger from the Stamp Act.[18] Many printers had difficulty sustaining their businesses because there simply was not much demand for their services in many small towns. Customers offered little help: they were slow in paying for their newspapers and, in the case of commercial advertisers, their advertisements.[19] In addition to these endemic problems, printers, like others, faced a general collapse of the American economy in the aftermath of the war. Colonists made money supplying the British Army during the Seven Years' War, which infused much-needed specie into the colonies. But depression set in with the advent of peace in 1763. Profit margins remained very low and customer payments erratic. As Parliament's solution to pay down its wartime debt, the Stamp Act thus placed printers at the center of a perfect economic storm. The requirement to pay the duty up front in sterling added further injury to the act's insult; printers rarely had enough money on hand to pay the duties even after they had sold their products, let alone beforehand. Many printers now found their trade highly precarious.

Faced with these prospects, some printers despaired. David Hall of Philadelphia, who had long taken a rather conservative stance on political issues, wrote numerous letters in 1765 to Benjamin Franklin and William Strahan in London. In the spring, Hall feared that he would have to give up his newspaper business altogether because, as a result of resistance to the act, "not one in Ten of our People" would be willing to purchase newspapers bearing the stamp.[20] As early as June, Hall worried that local residents' opposition to the Stamp Act would render the number of subscribers to the *Pennsylvania Gazette* "very trifling" and that the duty on advertisements would "knock off the greatest Part of them."[21] Unlike Strahan in London, he could not rely on street hawkers to sell excess newspapers "and pay the Ready Money for them" because most of his customers were spread out over a broad geographic area.[22] Having taken the measure of the people, Hall had determined that his business had no prospects for the future if the Stamp Act were to take effect.

Far from the growing militancy of the colonial crowds, both Franklin and Strahan gave advice that seemed to make sense in England but proved tin-eared in America. Each suggested to Hall that the Stamp Act and the resulting drop in business would allow Hall a counterintuitive method to strengthen his financial position. Both thought that the new requirement to have cash on hand meant that Hall should end his practice of accepting subscriptions and advertisements for the *Gazette* on credit. It was true, they admitted, that he would lose some customers, but that did not matter for two reasons. First, people would be so hungry for news that they would "soon return to their subscriptions," notwithstanding their protestations to the contrary.[23] Second, Hall's business would be far more solvent because the requirement to pay the duties in cash prior to sale allowed him to force customers to pay in advance.[24] Both thus dismissed Hall's on-the-ground estimation of the political fervor against the Stamp Act and ignored his pleas that the print market operated much differently in the diffusely settled American colonies than in metropolitan Britain. Strahan also posited several ways to treat the law as just another cost of doing business, such as by printing the advertisements separately and giving them away for free (a form of publication not covered by the law), a method that London printers had used to circumvent the 1712 stamp tax. And at worst, Strahan suggested, Hall could buy time by forcing local stamp officials to write back to England for instructions.[25] Long dyspeptic about his financial and political prospects, Hall remained unimpressed with these reassurances and proceeded with an abundance of caution in addressing the act and its consequences.[26]

Most printers, however, initially saw potential for profit in the controversy. As soon as the text reached North America, several printers reprinted the Stamp Act as a cheap pamphlet. For example, John Holt in New York and William Bradford in Philadelphia each sold an edition for one shilling.[27] Many printers published excerpts of the act related to the medium in which the excerpt appeared, most often newspapers or almanacs.[28] *Ames's Almanack*, which was published jointly by most of the printing offices in Boston, included an excerpt warning of the penalty for selling almanacs without stamps.[29] While printers mulled over their long-term strategy, they could at least bring in additional income.

Printers also used the coming of the Stamp Act as an advertising tool, emphasizing the urgency of purchasing their goods before the tax took effect on November 1. Because almanacs had a seasonal sales market, printers typically sold the following year's edition in the late fall—that is, just as the Stamp Act was scheduled to take effect. Such a sales pitch would also work well because almanacs were a popular steady seller in the American colonies, second only to the Bible. In addition, printers exerted extensive control over almanacs (more so than over any other form of media), often serving as author, editor, and compiler for each edition.[30] Printers who put the Stamp Act in their almanacs emphasized its inclusion as part of the almanac's function of serving the public and augmenting its general knowledge. When David Hall published *Poor Richard Improved* for 1766, he printed a letter (signed by the pseudonymous Richard Saunders) urging readers to familiarize themselves with the Stamp Act—the "Substance" of which was in the almanac—"in order to avoid the many Forfeitures they might otherwise be liable to."[31] When publishing their pamphlet edition of the Stamp Act in Boston, Edes and Gill must have found the language of Poor Richard particularly compelling. Less than three weeks after *Poor Richard Improved* appeared in Philadelphia, Edes and Gill used the same language in the *Boston Gazette* to try to persuade their readers to purchase the pamphlet.[32]

The Stamp Act forced several printers to move up their publication schedules for their almanacs in order to gain commercial advantage. Because the four-pence duty on almanacs was scheduled to take effect before most 1766 almanacs would have been sold, several publishers advanced their production forward into the summer of 1765. By August 1, William Bradford had already published the new *Pennsylvania Pocket Almanack* in Philadelphia, and William Goddard had also printed the *New-England Almanack* for 1766 in Providence, noting that it would be sold "*unstamp'd.*" John Holt published his new *Poor Roger's American Country Almanack* on September 20.[33] Many emphasized the savings to customers who bought their almanacs before the duty could take effect. Andrew Steuart, another Philadelphia printer, ran advertisements in Bradford's *Pennsylvania Journal* encouraging patrons to purchase their almanacs early, "as it will save the printer the trouble of getting them stamp'd, and the buyer at least six pence in each Almanack."[34] Whether buyers snatched up these almanacs early is unknown, but the message was simple: buy early or pay more later.

Printers also used their almanacs to make subtler protests. For example, the monthly calendars that were a staple of eighteenth-century almanacs typically featured lists of saints' days, birthdays of the royal family, and days of import in English history, such as the anniversary of Charles I's beheading. Two Philadelphia almanacs—the *Gentleman and Citizen's Pocket-Almanack*, published by Andrew Steuart, and the *Pennsylvania Pocket Almanack*, published by William Bradford—featured an altered notation for November 1. Instead of labeling the day "All Saints," as most other almanacs did, they denoted it as "All Slaves," a glib but effective use of the rhetoric of enslavement that was circulating throughout the British Atlantic seaboard that summer.[35]

Printers faced a more vexing dilemma as the implications of the act became clearer and newspaper subscribers began to declare their intent not to receive stamped newspapers. Whether to continue to print without stamps was a fraught decision for many printers, who mostly had not previously taken sides openly in major political controversies.[36] But Patriot groups in the thirteen contiguous mainland colonies had eliminated the option of printing *with* stamps by forcing the officers appointed to distribute stamps to resign. To print without stamps put a printer in violation of the law, and penalties were steep, up to twenty pounds per violation for using unstamped materials. The act imposed the same fine for printing and selling newspapers or pamphlets without printers' names and places of business.[37] Printers in most colonies were also caught between the branches of government. Most assemblies, elected by local colonists, opposed the act, whereas the governors supported the act in their roles as representatives of the empire. Because many printers relied on government work for the bulk of their income, choosing sides among parts of colonial governments became treacherous.

The Act Takes Effect: To Print or Not?

The printers' most important business decision in 1765—whether to print after November 1—was heavily mediated by politics. Printers reacted in various ways, but all of them decried their situation and dramatized the danger to their livelihoods. How printers reacted corresponded closely to their financial situations and the underlying strength of their businesses, even as they framed their actions in political terms. Their decisions are easiest to trace in the case of newspapers, which carried a date stamp and so can be checked for compliance. Printers had four choices with respect to their newspapers: they could publish with the stamps and pay the duty; cease publication; publish anonymously, that is, without an imprint; or publish openly without stamps in the face of potential sanction. Each carried financial risks, so many printers vacillated among several options at different points during the crisis.

Most printers who used the stamps worked in Canada or the West Indies, where the printing business relied more heavily on the good graces of the government for work and where popular protest never rose to the level of nullifying the act. Consequently, stamp officers were in place, and stamp paper was in ready supply.[38] Such printers typically published "Court papers" that served as the official organ of the colonial government. Extant copies of the *Barbados Gazette* and *Barbados Mercury*, for example, both bear stamps.[39] The group also included Anthony Henry, printer of the *Halifax Gazette*, and Alexander Shipton, publisher of the *Antigua Gazette*, who apparently printed his newspaper through November and December 1765 with stamps, until the stamp officer there resigned and declined to provide him additional stamps. When American papers published an excerpt from the *Antigua Gazette* announcing that Shipton could no longer use stamps, they noted that the newspaper was not originally printed "*on Stamp Paper*."[40] The group that used stamped paper to publish its newspapers and almanacs was small, however, both because opposition among colonists as a whole was so strong and because there were so few government printers on the mainland who did not rely on the colonial assemblies, where most of the official protests against the Stamp Act originated.

A second group of printers—those who refused to print and temporarily closed their shops—tended to face substantial underlying business woes. Less than a month after news of the act had arrived in America, William Goddard announced the suspension in publication of the *Providence Gazette*, specifically citing the "oppressive and insupportable STAMP-DUTIES" as the reason for his decision.[41] The *Gazette* had been struggling financially for some time, and Goddard took the opportunity to go to New York to work as a journeyman for John Holt. However, Goddard thus ceded (for the most part) the opportunity to shape debate in Rhode Island and elsewhere. Further south, James Parker abandoned his plan to begin a newspaper in Woodbridge, New Jersey, writing to his patron Franklin that "the News of the killing Stamp, has struck a deadly Blow to all my Hopes on that Head."[42] Notably morose about his business prospects throughout his career, Parker felt blocked from one of the most significant paths to establishing a reputation as a printer. Newspaper publication may not have been particularly lucrative, but it was a steady business and placed the printer's skills before the public. Yet Parker's prospects were never particularly rosy, for he was at the same time waging financial battles with not one but two former partners in New York, William Weyman and John Holt.[43]

Whatever their economic woes, printers who suspended their newspapers blamed the politics of the Stamp Act for their difficulties. Jonas Green and William Rind, for example, who published the *Maryland Gazette*, suspended its publication with the issue of October 10, 1765. That date happened to coincide with the anniversary of their partnership and thus provided the opportunity for a clean break for their joint business relationship, as Rind planned a move to Williamsburg. Yet in the newspaper Green heralded the end of its run as a political event. He anthropomorphized the paper, decrying its demise in terms ordinarily used for the death of a notable. For the final number, Green utilized many of the tools of his trade. He draped the front page in mourning borders, a practice typically used only for the death of a grand personage, such as when King George II died in 1760. Beneath the masthead, he included an epitaph for the newspaper: "E X P I R I N G :/In uncertain Hopes of a Resurrection to LIFE again"—an epitaph that other printers, including William Bradford, would borrow later that October. The pithy formulation summed up well the cloudy future that printers foresaw, while leaving open the possibility that the newspaper might one day return. He also published an address to readers, seeking final payment from delinquent subscribers, and alluding (in sharply contrasting Fraktur type) to "that **Dooms-Day**" and "that **dismal Day**," without ever referring to the Stamp Act by name.[44] Boston newspapers picked up on the metaphor, announcing the end of the *Gazette* in the language of an obituary.[45]

A third group of printers attempted a subtler subterfuge, continuing to print after November 1 but without attaching their names or newspaper titles to their publications. These printers placed their political interest in opposing the act (and, in a few cases, avoiding the crowds) ahead of any repercussions they might face from government sanction if caught printing illegally under the act's terms. Anonymous publication occurred most prominently in New York and Philadelphia, where printers of various backgrounds, including Sons of Liberty members like John Holt, and more conservative, older printers such as David Hall and Hugh Gaine, all published short, single-sheet accounts of news in the first weeks of November. The news sheets typically contained a brief title as a masthead in place of the newspaper's title. For example, several printers, including David Hall, published sheets with the header "No *Stamped Paper* to be had" or the phrase "Remarkable *Occurrences*" as the masthead.[46] Most of them contained no advertising whatsoever, a significant financial hit for printers. Because of the strength of anti–Stamp Act sentiment, continuing to print, even in an imperfect way, was construed as an act of patriotism. By the middle of November, though, most of these printers had once again begun to print their newspapers under their own names and under the full mastheads of their newspapers. For them the fears of suffering British penalties took second place to either the genuine desire to resist more openly or the risks of appearing insufficiently patriotic. These printers, who published in defiance of the law, continued to face difficult economic conditions: advertising, for example, continued to lag through the rest of 1765 in part because many merchants had agreed not to purchase goods from England until the Stamp Act was repealed, diminishing their need to advertise for the time being.

Some older, more established printers tried to take a conservative tack to the protests by continuing to publish without stamps, but they did so only upon the exertion of political pressure. Such men were generally unwilling to put themselves on the line politically in order to effect change and so found themselves the frequent target of pressure by the Sons of Liberty and other groups opposed to the Stamp Act. Included in this group were longtime associates of Benjamin Franklin—members of his printing network, such as James Parker in New York, David Hall in Philadelphia, and Jonas Green in Annapolis. Many of those who acted only after pressure had roots in the British Isles. Hall was a Scot, as was James Johnston of Savannah, Alexander Purdie of Williamsburg, and Robert Wells of Charleston. A few had been born in England, such as Joseph Royle of Williamsburg, and one, Hugh Gaine, was an Irishman. This group of printers—whom William Goddard later characterized as "cautious & opulent Men"—more often suspended their papers as they tried to navigate between their fears of the law and the mob.[47] They also tended to avoid controversy. Royle, for example, declined to publish the radical Virginia Resolves, proposed at the end of the session of the House of Burgesses in May 1765 by a young firebrand lawyer named Patrick Henry.[48] Gaine, among others, published essays that argued both sides of the question of whether the Stamp Act constituted a legitimate tax.[49]

A final group of printers resided almost entirely in Boston, where printing openly, that is, under one's own name, and without stamps was most common. Opposition to the Stamp Act had been pervasive and highly visible in Massachusetts, which had led the way in protesting the act and forcing its appointed stamp officers to resign over the summer. Printers therefore had more leeway to continue publishing, sensing that they were less likely to be penalized. Benjamin Edes and John Gill were the boldest: they simply continued to print the *Boston Gazette* under its usual masthead and with their names listed as if nothing exceptional was afoot. Edes was a member of the Loyal Nine, the core group of the Boston Sons of Liberty, and so had been particularly active in opposing the new law. Other printers were less militant, but all the Boston newspapers continued to publish through the entire crisis. The Drapers, for example, made minor adjustments to the title in their *Massachusetts Gazette. And Boston News-Letter*. They shortened it to the *Massachusetts Gazette* and printed each issue as "No. 0" in protest. The *Boston Evening-Post*, published by the brothers Thomas and John Fleet, refrained from printing their names for six months, from November 1765 through the following May, but continued to publish.[50]

Compared to their Boston compatriots, printers elsewhere employed a broad range of methods to demonstrate their objections. Some protested in a simple way, printing or reprinting articles in their newspapers about the Stamp Act and then refusing to print or, in some cases, insisting on printing without stamps. Some printed additional pamphlets on the Stamp Act or such shorter works as "Oppression: A Poem," a twenty-page diatribe against the ministry originally published in London, which was for sale at the printing offices of Hugh Gaine in New York, William Bradford in Philadelphia, and Edes and Gill in Boston.[51] Several printers, like Edes, actively participated in the Sons of Liberty in their towns. William Bradford took a leadership role with the Philadelphia Sons of Liberty, including acting as one of its correspondents and signing several letters for the group.[52] William Goddard was active in Providence, and John Holt was closely associated with the group in New York.[53] These printers stood out, of course, and their compatriots who lack a full documentary record likewise reacted to the crisis.

During the crisis, printers sought to merge their political and commercial interests as best they could. Printers initially began preparing for the Stamp Act to take effect, planning how to accommodate the price increase and attempting to keep their readers abreast of how the price structure for their purchases would change. Once it became clear, however, that the Sons of Liberty had in practice nullified the Stamp Act, printers faced the difficult choice of passively siding with the law by refraining from printing—but potentially incurring the wrath of protesters—or siding with the protesters and continuing to print at the risk of their offices and livelihoods should the imperial administration succeed in enforcing the law and pursuing prosecutions against the printers. Even printers who continued to publish stood to lose revenue from slackening sales of their newspapers and almanacs, reduced advertising, and the concomitant drop in business for their ancillary interests in stationery and other goods. Though it would be impossible to quantify precisely because of a dearth of records, printers suffered greatly through the fall and winter of 1765 and 1766.

External Political Pressures and Influences

Many printers acted more boldly in 1765 than they previously had to protest the Stamp Act, but all still had to balance their own political views, their commercial interests, and the political winds that surrounded them. Like their fellow colonists, as individuals they held a diverse range of political views, whether they expressed them publicly or not. Their decisions about whether to expose those views in print, however, carried implications not just for themselves but also for the entire community. For that reason, printers were not the only ones interested in the political perspective offered up in their publications. Nor could they make decisions independent of prevailing attitudes and their own commercial prospects. During the crisis—and continuing afterward—politically active men and organizations offered sustained attention to the press as a vehicle for political ideas and sought to exercise influence on how printers portrayed events, what essays they published, and so on. Partisans did so through economic means such as subsidies or boycotts and sometimes resorted to more forceful and violent measures when they felt it necessary.

During the months-long crisis, printers engaged in a delicate negotiation with extralegal groups over how newspapers and other forms of print should serve anti-imperial protest. This occurred whether or not the printers themselves were members of the extralegal groups. The Sons of Liberty, the most prominent of these groups, for the most part had very little interest in the financial implications of the Stamp Act for printers or in the possible penalties printers might face should British authorities successfully enforce the law. However, the group offered support to some printers, as when it paid a £440 legal judgment against John Holt in the fall of 1765.[54] Its political aim was to oppose the Stamp Act and prevent its taking effect. It leveraged that goal into support for printers whose views aligned with its own and created trouble for printers whose public stance it deemed too cautious.

The Sons of Liberty and its ideological allies exerted strong pressure on printers who remained neutral or pro-government to continue publishing and to print material in line with its political views. In Massachusetts, it turned eighteenth-century journalistic conventions against Richard and Samuel Draper, publishers of the *Massachusetts Gazette. And Boston News-Letter*. Most newspapers during the colonial period published relatively little local news, opting instead to serve as a "journal of occurrences" of British and European news.[55] This practice rang false, however, in the wake of the riots in Boston on August 14 and the destruction of the house of Andrew Oliver, the Massachusetts stamp officer. The Drapers declined to publish an account of events that had occurred in their own town, for which an anonymous writer took them to task in the rival *Boston Evening-Post*, published by Thomas and John Fleet. "A.Z." praised the Fleets effusively, arguing that they were known "*to be too well-affected to the Common Cause of Liberty*" to fail to publish "*some late* important Domestic Occurrences," as "*the Printers of Thursday's Paper did* [i.e., the Drapers] *tho' they gave an* extraordinary *Half Sheet full of* immaterial Foreign *Advices*." The anonymous author presented the Fleets with his own account of the riots, a "*plain Detail of Facts, without any Exaggeration*," and urged that the Fleets should print it, "*unless to make Room for some better Account*."[56] They did, of course, and the account proved very influential. Within a few weeks printers in several colonies reprinted it, though they left off the preface, which had only local significance.[57] That the Fleets published A.Z.'s preface suggests that they agreed with his insinuations that the Drapers had political motives for failing to report on the riots. The printing trade was not zero-sum, but it was competitive in Boston, so the Fleets also may have hoped to gain some commercial as well as political advantage by pointing out the Drapers' omission.

The Sons of Liberty encouraged public economic pressure on other wavering printers. Peter Timothy, a Charleston printer and longtime associate of Benjamin Franklin, became the target of protest by his newspaper subscribers after rumors began to circulate that he had been appointed the stamp distributor for South Carolina. Such an appointment would not have been a surprise given his connection to Franklin, who had already secured him a position in the imperial post office.[58] Timothy had also publicly prepared his customers for a possible price increase because of the act. He promised that any "*addition* will be as moderate as possible" and committed himself to supplying his customers with his *South-Carolina Gazette* after November 1 unless they indicated otherwise.[59] Just before November 1, 1765, however, Timothy decided to end the run of the *Gazette*, after finding that "the numerous subscribers have signified, *almost to a man*, that they will not take in ONE stampt news-paper, if stamps could be obtained." Timothy, among the least financially viable printers in the colonies, did not publish his newspaper again until June 1766.[60] Finding himself "reduced to the most *unpopular* Man in the Province," he felt obliged to refute the rumor that was "*industriously* and . . . *maliciously* circulating" that he had been appointed as stamp commissioner. He categorically denied that he had taken up the commission, had sought appointment, or knew of anyone who had sought appointment on his behalf.[61] Though Timothy would later become a staunch Patriot, even the hint of acquiescence to the act precipitated massive commercial trouble for him.

Even when printers were supportive of the Sons of Liberty or directly involved with the group, they could still face intimidation. Take, for instance, the case of John Holt, who was active with the group in New York and printed numerous anti–Stamp Act essays and pieces in his *New-York Gazette*. When November 1 arrived, the Sons of Liberty, which had effectively nullified the act in New York, appeared curiously anxious about whether Holt would back up his support by continuing to print the *Gazette* on unstamped paper. To ensure his cooperation, it ordered him to do so by sending a letter to his house and posting it publicly at the nearby coffeehouse. The letter, which was signed by "John Hamden" on behalf of "*a great Number of the free-born Sons of New-York*," opened with the famous epigram of Horace, "Dulce et decorum, pro Patria mori" (It is sweet and proper to die for one's country).[62] Hamden praised Holt as "*a Friend to Liberty*" who had published "*such Compositions as had a Tendency to promote the Cause*" but expressed fear that Holt might "*shut up the Press, and basely desert us*," in which case, it warned, "*depend upon it, your House, Person and Effects, will be in imminent Danger*."[63] What makes the case especially interesting is that an account of the letter's public posting was printed in Philadelphia by David Hall, who up to that point had attempted to remain neutral in the crisis. That he chose to print an unflattering piece about Holt may reflect Hall's connection through the Franklin network to James Parker, Holt's onetime partner in New York, who was at the time in litigation against him over the profits of their former partnership.[64]

In dealing with the crisis, printers also faced pressure from royal officials, many of whom heaped blame on the printers themselves. For example, Massachusetts governor Francis Bernard reported to the Board of Trade that "the infamous set of Newspapers" published during the crisis proved "the immense pains [that] have been taken to poison the Minds of the People."[65] In a few cases, imperial officials threatened or harassed printers. In March 1766, for example, a Lieutenant Hallam, stationed on the *Garland* in New York harbor, allegedly threatened John Holt with physical harm for materials he had published in the *New-York Gazette.* According to depositions taken by the Sons of Liberty, Hallam "declared, that what was printed in the Thursday's Paper here, was equal to the proceedings in Scotland in the Year 1745," referring to the Jacobite rebellion against the British Crown. Upon further examination, Hallam argued that "had a Printer in England done as much, he would have been hang'd," and suggested that "he himself would not be too good to put a Rope about his Neck."[66] The Sons of Liberty threatened to retaliate against Hallam, who took refuge aboard his ship. The story spread to Philadelphia and Boston before it disappeared from the news.[67] For the most part, however, imperial officials vented their frustration in letters to London, while not taking action against colonial printers.

Circulation Patterns and Protest Strategies

Printers not only had to decide whether to continue printing their periodicals and other publications; they also had to determine how to portray the events going on around them. Arriving in their printing offices daily was a plethora of printed and unpublished written materials, oral reports, images, suggestions, threats, and rumors. From this array each printer made choices about what he should publish. During the Stamp Act crisis, a variety of textual strategies played important roles in how the crisis was projected in print. Printers wove narratives that aimed to shape public opinion. They might portray protests as theatrical performances devoid of violence and full of unity (or the opposite, if the printer was a "friend of government"). Some printers took advantage of trade conventions—sometimes as seemingly pedestrian as the use of personal pronouns—to elide or enhance the origins of an account. And then they circulated their publications for their brethren to reprint, replicating the process in dozens of offices throughout North America.

One of the greatest opportunities for printers to influence perceptions was in chronicling the riots against the Stamp Act in Boston and other towns in the late summer and fall of 1765, actions that themselves possessed great symbolic import. These crowd actions fell into a pattern common in eighteenth-century America and England. Before the Revolutionary era, riots and demonstrations became more common, especially in growing urban areas. They might arise in response to the possibility of slave revolts, disputes over smallpox inoculations, and other threats to the accepted order. Not all crowd actions protested local social hierarchies. Instead, many (especially in larger towns) favored the protection of community customs and cultural norms against the encroachment of central administrative authorities.[68] Furthermore, government and social leaders often mobilized crowds for their own purposes, particularly in punishing criminals. During the eighteenth century, open-air punishments served as a means of communicating the power of the state to regulate and punish: the crowd legitimated the process of confession and execution. These events, whether they entailed a state-sanctioned public execution or a "bottom-up" demonstration against small pox inoculation, all contained elements of the theatrical. That is, by the eighteenth century the political actors involved in these controversies were aware that their actions had symbolic weight.[69] The hanging in effigy of a royal official and his subsequent "execution" by fire was a clever and very visible inversion of the power of the state to dispense punishment. The crowd itself, sometimes in alliance with local officials, claimed power to enforce its rights by symbolically punishing those who made and executed unjust laws.

The Boston Stamp Act riots provide a useful illustration of this phenomenon. By August 1765, tensions had risen in Boston over the news that Andrew Oliver, brother of Chief Justice Peter Oliver, had been appointed the Massachusetts stamp collector. On August 14, a mob, possibly instigated by the Loyal Nine and led by South End shoemaker Ebenezer McIntosh, hung effigies of Oliver and Lieutenant Governor Thomas Hutchinson, marched to Oliver's home, and demanded that he resign. Having never received an official commission, Oliver attempted to satisfy the mob by announcing rather vaguely that he would take no action to enforce the act, but, far from being appeased, the crowd ransacked his home. Less than two weeks later, on August 26, another mob formed, storming Hutchinson's new home and causing considerable damage.[70]

Reaction to the two Boston riots differed markedly. In the case of the first crowd action on August 14, local elites supported it as a legitimate way for colonists to assert their constitutional rights. The Massachusetts House of Representatives had sent a petition to Parliament and King George III, but their remonstrance did nothing to prevent the law from being enacted. The only remaining recourse both elites and ordinary townspeople saw was to nullify the law through crowd action.[71] Targeting Andrew Oliver, the appointed stamp officer for the colony, the actions of August 14—including the symbolic execution and pressure exerted by threats to Oliver's personal property—fit well-established patterns of ritualized violence. The second riot, on August 26, by contrast, did not have the support of elites, mostly because it targeted other colonial officials not directly involved in the administration of the Stamp Act, Thomas Hutchinson in particular. The second riot therefore fell outside the bounds of legitimate crowd violence.

The riots in various North American towns during the late summer months were highly stylized, and printers augmented the theatricality of their reporting. All the riots took place in urban settings or larger towns primarily along the Atlantic seaboard and in locations that roughly corresponded with the sites of printing presses. The riots therefore occurred in the context of continuous communication in print among towns along the coastline. News was then broadcast into the interior through the port town newspapers, many of which had numerous subscribers in the hinterlands. Printers and other authors framed and transmitted their accounts in ways that would make distant readers feel as if they were eyewitnesses to the riots.

The Stamp Act riots dominated American news in the weekly newspapers. In just one issue of the *Boston Evening-Post*, for example, the Fleet brothers printed Stamp Act–related news from a range of colonies, including the two proclamations issued by Governor Bernard on the Boston riots of August 14 and 26; the proceedings of a town meeting in Boston condemning the second riot; news of riots in New London, Norwich, Newport, and Providence; accounts of the forced resignations of stamp officers in Virginia, Connecticut, New Hampshire, and New York; and a list of the items stolen from Thomas Hutchinson's home on August 26.[72] Collectively, these items dominated the newspaper. This was especially notable because most newspapers before 1765 had focused primarily on European news, providing news from other colonies only in small doses. By turning their attention to other colonies, printers insinuated that their readers should note well the occurrences not only in Europe but also in North America.

Some items about the Stamp Act riots were brief accounts of events, as when several newspapers reprinted a report from Annapolis that a group in Baltimore had "diverted themselves with Carting, Whipping, Hanging, and Burning the Effigies of a Distributor of Stamps."[73] Others shared characteristic patterns with reporting on eighteenth-century rituals of public punishment and popular protest. These accounts drew the reader into the frame of the story and related with great specificity the actions of the crowd.[74] In structuring reports of the Stamp Act protests, chroniclers organized the sequence of events into an ordered narrative. First, someone would publicly display an effigy or series of effigies in a public square, almost always of a stamp officer (generic if not the colony's officer), and occasionally one of George Grenville or Lord Bute. The effigies were decorated with poems or given tags quoting their "thoughts" as they hung on public display. Come evening, the crowd would gather to rally against the Stamp Act and occasionally parade the effigies through the town before "executing" them in a bonfire.

The Stamp Act riots occurred in rapid succession in numerous towns in late August, too rapidly for news of the first riot in Boston to have reached all of the places that subsequently rioted. Yet the narratives of the events elsewhere closely resembled the early reports from Boston, largely because the authors and editors understood the formula for setting the scene of such riots. Many appear to have been designed as much for consumption by readers in the hinterland and in other colonies as for consumption by the local readers who may have witnessed the action in the towns where the protests took place. Thus, the reporting of events was affected by how they would appear to readers elsewhere in the colonies. Newspapers were usually responsible for the first written accounts but not always. Sometimes printers received personal letters from eyewitnesses. Either way, newspaper publication was a central mechanism for moving the event out of the realm of a local and transient experience into manuscript and print for circulation into other regions.

For example, a letter from Boston printed in the *New-York Mercury* and *New-York Gazette* recounted the discovery of an effigy of Andrew Oliver on August 14. The letter writer was careful to note that the effigy "was found hanging on one of the Trees at the South End of the town," absolving any individual or group from responsibility. As if the reader were standing with the writer, he described what was written on the effigy, a text accusing Oliver of betraying his country, at the expense of his soul:

> Fair Freedom's glorious Cause I've meanly quitted
> ------ -------- ----- For the Sake of Pelf,
> But ah ! the Devil has me outwitted,
> And instead of stamping others I've hang'd myself.

In a burst of rhetorical enthusiasm, the effigy makers associated their protest with freedom and identified it as a "glorious Cause," at the same time implying that Oliver made a deal with the devil for profit. Even though the stamp officers' identities were already common knowledge throughout the colonies, the writer noted that the effigy was identified simply as "A.O." on his right sleeve and emphasized that the paper on the effigy warned, "Whoever takes this down, is an Enemy to his Country."[75]

A few weeks later, the *Boston Evening-Post* and *Boston Gazette* reprinted an "Extract of a Letter from a Gentleman" in Newport reporting to a friend in Boston on the events of August 29. The author recounted how "*the populace*" in Newport had "*brought forth the effigies of three persons in a cart*" and displayed them on a twenty-foot-high gallows in the center of town. As with the Boston riots, the effigies were identified as the stamp officer and two of his fiercest allies, and they were marked "*from his pamphlets* [with] *several motto's*" to explicate their political stance against the people. One of the posts was topped with lyrics to a song, an important genre in "playing" at rebellion and making it plausible.[76] The lyrics harangued those who would "betray" their native land "for a Post, or base sordid Pelf," and urged patriots to "stand strong . . . / To maintain our just Rights," so that "These Effigies first–next the Stamp Paper burn." Just as in Boston, the mob was bent on destruction, yet here too elites interceded and were able to head participants off from ravaging the stamp distributor's home by promising to "*deliver up Mr.* Johnston*'s effects the next day, . . . unless he resigned the Office*."[77] The gentleman made clear that he found the crowd's protest to be despicable. However, in hitting each of the points of the genre of writing about crowd participation, the anonymous author implied that the crowd's actions were legitimate.

Several accounts of the Boston riots competed in print for the attention of printers and their readers. The first to appear was a proclamation by Governor Francis Bernard seeking information that would lead to the arrest of the August 14 rioters, but it was printed only in the *Massachusetts Gazette* and not until after the August 26 riot, when the other Boston newspapers printed that proclamation together with a separate one issued after the second riot. The *Boston Gazette*'s account, which Edmund and Helen Morgan claimed was reprinted "from one end of the continent to the other," did not actually circulate that widely—only to the *Providence Gazette* and the *Newport Mercury*.[78] Instead, most colonial newspapers republished the account printed by Thomas and John Fleet in the *Boston Evening-Post*.[79] That printers chose to reprint the Fleets' account rather than that of Edes and Gill—who were well connected politically to the Sons of Liberty and the elite Boston leadership—seems to indicate the lack of intercolonial political cooperation at this early date rather than a strong network centered around the Fleets. Not until very late in 1765 and into 1766 did groups styling themselves the Sons of Liberty begin to organize across colonial borders.[80]

Printers also shaped the tone of political discourse by transforming and editing accounts of the riots. Bernard's official proclamation, for example, saw very little circulation—except, somewhat incongruously, in the pages of the *New-York Gazette*, published by Sons of Liberty member John Holt.[81] When he reprinted the account, however, Holt gave no indication that he was using a government proclamation as his source. Instead, he used an impersonal voice common in newspapers, opening the paragraph with the preface, "We find by the Boston Papers." This rhetorical maneuver thus elided—in fact erased—the point of view of the original account. A reader of the *Gazette* had no way to know that the text he was reading was that of an imperial official whose first response to the riot was outrage at activities that defied authority. Holt asserted great power as an interpreter of Boston events to judge what was fit to print for his readership. He in fact accentuated Bernard's emphasis on the unruliness of the crowd by adding a bracketed postscript: "We hear, among other Feats they roll'd out 2 Pipes of Wine, which setting an End, and beating the Heads out, they regaled with, and emptied." In these ways, Holt interjected his editorial voice and exerted control over the circulation of information.

Other printers also used the authority conferred by the first-person plural, which was a standard journalistic device of the period to control the flow and context of information and reporting. Richard and Samuel Draper, two printers quite friendly to the government, apparently found inadequate the various accounts of the August riots and openly solicited a narrative of events from their readers. Yet, even as they opened their newspaper to outside voices, they kept a firm grip on how the events should be interpreted, setting the tone for what they expected the piece to say. In their solicitation, they wrote that they sought someone who could "*represent (if possible) the shocking Transactions*" of the night of the second riot and asked that the author reflect on the "*terrible Effects*" on the "*immediate Sufferers*" and "*the Community in general*."[82] In phrasing their request in this manner, the Drapers largely predetermined who would submit accounts and how the paper would portray the events of that evening.

Another printer, William Goddard, took the assertion of editorial authority even further by creating a fictitious persona under whose name he published some of the most vitriolic essays against the Stamp Act. In September 1765 he snuck across the Hudson River from New York to James Parker's new printing office in Woodbridge, New Jersey—without Parker's knowledge and possibly at the suggestion of John Holt, for whom Goddard was working at the time. On Parker's press, he published (anonymously) the *Constitutional Courant*, which he pitched as a new newspaper "Printed by ANDREW MARVEL, at the Sign of *the Bribe refused*, on *Constitution Hill, North-America*," with a header: "Containing Matters interesting to LIBERTY, and no wise repugnant to LOYALTY."[83] The masthead also included the "Join or Die" snake device, created by Franklin in the 1750s to emphasize the importance of unity in confronting the French and Indians during the Seven Years' War. The paper included an address to the public from "Andrew Marvel," and two essays signed by "Philoleutherus" and "Philo Patriæ" opposing the Stamp Act more openly than almost any other piece to date.

Marvell was an odd yet apt choice for a pseudonym in revolutionary America. Although a prominent poet and republican in Restoration England, his texts did not circulate widely in North America in the eighteenth century.[84] However, in August 1765 a letter appeared in the *New-York Mercury*, published by Hugh Gaine, purporting to tell the tale of Andrew Marvell, who defended liberty against the "wicked" ministers of the king at the expense of his own well-being. According to the letter, signed under the pseudonym "Hampden Sidney" (thus invoking two other seventeenth-century republicans near and dear to colonial hearts), Marvell was elected to Parliament by the Borough of Hull, and "His Understanding, Integrity, and Spirit, were dreadful to the *then* infamous administration" of King Charles II. The ministry thought him "theirs for properly asking" and dispatched his onetime classmate, the Lord Treasurer, to dine with him and thereupon offer Marvell a thousand-pound note—the bribe that Marvell refused.[85]

Goddard, under the guise of Marvell, used the standard device of the editor's address to the public to attract attention to these political essays. Such addresses appeared in the first numbers of many eighteenth-century newspapers and typically made straightforward claims. They committed the printer to operating a free press, that is, one open to all parties, and to publishing the newspaper regularly, usually weekly. They also typically announced that the printer was available to do all manner of printing jobs. In the *Constitutional Courant*, "Andrew Marvel" told the story of how these essays came to be published. "Marvel" wrote that the authors of the essays, "Philoleutherus" and "Philo Patriæ," had attempted to get them published in New York, only to be turned down by the printers there, whose friends advised them "to be careful not to publish any thing that might give the enemies of liberty an advantage." The essays, according to Marvel, were "thought to be wrote with greater freedom than any thing that has yet appeared in the public prints." Marvel was willing to concede the point but reversed his evaluation of the result. Unlike his colleagues in New York, "I, who am under no fear of disobliging either friends or enemies, was pleased with the opportunity of turning my private amusement to the public good." By cloaking himself in the pseudonym of a long-dead patriot and defender of freedom, Goddard gave himself cover to act in the public interest. "Andrew Marvel" also announced that he would "now inform my countrymen, that I shall occasionally publish any thing else that falls in my way, which appears to me to be calculated to promote the cause of liberty, of virtue, of religion and my country, of love and reverence to its laws and constitution, and unshaken loyalty to the King."[86] In closing the address, Goddard struck every patriotic note he could muster, all the while attempting to link virtue, liberty, and loyalty to the King to biting criticism of the King's ministers.

This single-issue newspaper immediately became a sensation. It appeared on the streets of New York on September 21, where, according to Goddard, "the sale was rapid & extensive."[87] Within the next few weeks, extracts were reprinted in several other colonial newspapers. In October, for example, an excerpt of one of the essays appeared in the *Newport Mercury* and the *New-London Gazette*.[88] The *Boston Evening-Post* republished Andrew Marvel's address with the "Join or Die image" and noted that "there is such a Demand for the above mentioned Paper in these parts, that, we hear, it will soon be re-published."[89] Discussion of the *Courant* reemerged in January when American newspapers reprinted excerpts from a London paper that noted that the *Courant* issue had reached England and caused a stir there—a perfect example of the feedback loops that dominated the transatlantic transmission of news. The London newspaper also carefully described the essays as "merely loyal."[90] Andrew Marvell (the historical figure) would make another appearance in American newspapers in 1766. The editors of the *New-Hampshire Gazette* cited Marvell's famous statement about the ability of the press to flush out corruption to defend their decision to publish a series of letters from John Hughes, the Pennsylvania stamp distributor, to his friend Benjamin Franklin. Hughes claimed that his letters were confidential and had sued William Bradford, the Philadelphia printer who first published them.[91] He did so with good reason: in the letters, Hughes expressed the hope that the Stamp Act would be enforced.

Printers used their editorial authority to position themselves as arbiters of political debates. In doing so, they navigated the complicated intersection of public and private communication. In December 1765 Benjamin Edes and John Gill published a letter addressed to them in the *Boston Gazette*, inquiring whether Andrew Oliver planned to act as stamp officer for Massachusetts. The author alleged that all the other stamp commissioners in North America had already resigned, "some of them in a genteel and generous Manner." Oliver was the exception. In response, Edes and Gill visited Oliver, as they "*thought it prudent to acquaint Mr.* Oliver *with the Contents*" of the letter before they published it. As he had done in August, Oliver gave a measured and ambiguous response. He asserted, according to Edes and Gill, that though he had received a commission as stamp distributor, "*He had taken no Measures to qualify himself for the Office, nor had he any Thoughts of doing it.*" He understood the potential value of publicizing his statement, because he told the two printers that they had "*Liberty to assure the Public that he would not*" assume the office.[92]

Oliver did not state definitively that he was resigning, which proved insufficient to satisfy the Sons of Liberty. The following week's *Gazette* included a follow-up to Oliver's statement, in which "The True-born Sons of Liberty" aimed to force Oliver into a full public declaration of his intentions. The Sons of Liberty demanded that he appear "under the Liberty Tree" to resign publicly as stamp distributor. To buttress the request, the *Gazette* published the full text of "an Advertisement" that had already been "posted up in the most public Part of this Town," which invited the Sons of Liberty to meet Oliver at noon to receive his resignation. (Of course, the printed "advertisement" had been planted by the Sons of Liberty itself.) In an attempt to save face, Oliver tried to renounce the office at the Town House in front of a select group of elites. The crowd assembled at the Liberty Tree refused to accept this resignation and forced him to appear before the crowd to declare his intention not to serve as a stamp officer, a declaration witnessed by Richard Dana, a justice of the peace—another piece of theater.[93] Edes and Gill and the Sons of Liberty controlled the entire process, forcing the resignation to occur in public before a cross-class crowd rather than before those Oliver himself chose. For them, the publicity of the resignation, rather than the mere fact of it, was the true measure of success.

One of the most fascinating aspects of the protests appearing in print was the anthropomorphizing of the stamp paper itself. In numerous accounts of the protests, the paper was the featured villain. Because the act had taken effect in some parts of the British Atlantic, examples of stamped paper—Mediterranean passes (which allowed ships free traverse of the seas), customs forms, and newspapers—began to show up in colonial ports. The appearance of such papers often occasioned protest. The paper would be paraded through the streets, hung on a gallows, and then burned, as the effigies of administration and stamp officials had been. A February 1766 account in the *Boston Gazette* captured in tongue-in-cheek fashion the legal formalities to which colonists held themselves in "trying" an offending paper in "court." The paper, which was described as a "Prisoner" of the Sons of Liberty, was put on trial because it "did . . . endeavour to make its Appearance in a forcible Manner, and in Defiance of the known and establish'd Laws of the British Constitution, to deprive the Subject of his Rights and Privileges, &c." After a two-hour trial, a jury found the paper guilty of violating the Magna Carta and attempting to "subvert the British Constitution." Its punishment was to be burned along with effigies of Grenville and Bute.[94]

Printers also used their technological skills to enhance their protests. Edes and Gill, the printers of the *Boston Gazette*, rarely used images in their newspaper. The masthead did not contain one, and elsewhere images typically occurred only in advertisements, as was true of other newspapers at the time.[95] In this case, however, Edes and Gill provided a woodcut to accompany the description of the captured paper that graphically captured the hanging of effigies (fig. 4). The image brought to life the rich descriptions of the riots from the previous fall. The two figures in the woodcut represented George Grenville, architect of the Stamp Act, and Lord Bute, the king's favorite adviser. The Devil appeared above, lowering a paper to the shackled government ministers. The description re-created a literal stage, so that the reader became the bystander on the Boston street where the gallows was erected:

> A Stage having two Effigies thereon was erected, one of them representing B—te dress'd in Plaid, the other G——le; over whom was a Gallows, on which the Devil appeared with a Stamp Act, and a Stamp Paper in one Hand, and a Chain in the other hanging over the Gallows, with the following Words proceeding out of his Mouth to G——le, who held a Lock in his left Hand, *Force it*——to which he answer'd, *That we will upon the REBELS;*—B—te who held a Key in his right Hand replies; *We can't do it.!*——Upon the Stamp Paper were these Words, *For the Oppression of the* WIDOW *and* FATHERLESS.[96]

Edes and Gill varied their typography to emphasize key words and phrases, placing some in italics and using small capital letters to call attention to the oppression of the "WIDOW" and "FATHERLESS"—a phrase well known to colonists familiar with Scripture.[97] Most of the effigies hung in public contained some sort of written statement on them, in this case a metaphorical lock and key that "Grenville" and "Bute" debated using on the colonists. The stamp paper itself announced its purpose as the "oppression" of the weakest elements in society—not exactly how Parliament would have described a tax directed primarily at tradesmen, merchants, and lawyers.

Images played a crucial role in newspaper accounts of protest, in particular by deploying traditional symbols of mourning for use in political protest. In the days leading up to November 1, when the act went into effect, several newspapers used images of coffins or skull-and-crossbones designs to depict their demise and the arrival of the stamps. William Bradford, for example, loaded the front page of the *Pennsylvania Journal* on October 31, 1765, with doleful reminders of the day the Stamp Act was to take effect. He himself had published the resignation of the Pennsylvania stamp officer some three weeks earlier, so he knew that the law would be rendered moot in practice. The front page of the issue was lined with mourning borders, similar to the ones used by Jonas Green in the *Maryland Gazette* when he shut down that newspaper. Along the right side of the front page, written sideways, was the lament, "Adieu, Adieu to the LIBERTY of the *PRESS*." The image implicitly accused the British ministry of deliberately, and for political reasons, using oppressive taxes to shut down the American press. In the bottom right corner of the page, in the spot where according to the law a stamp must appear beginning the next day, Bradford published a skull-and-crossbones design (fig. 5). As if this imagery were not enough, Bradford on the last page of the

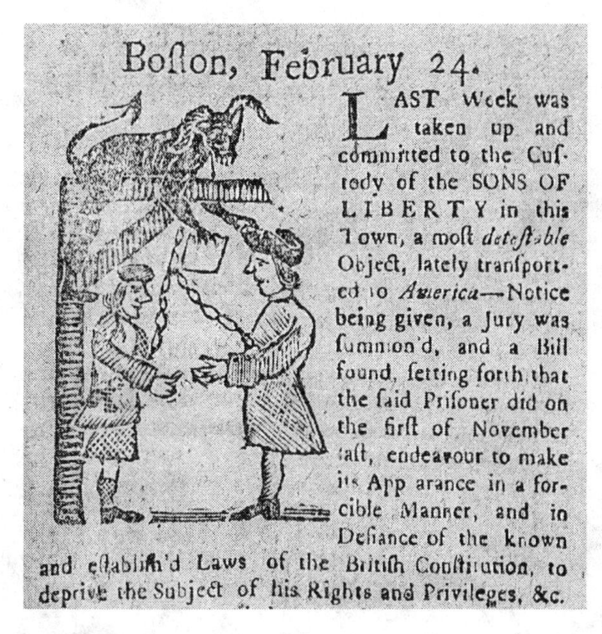

Boſton, February 24.

LAST Week was taken up and committed to the Cuſtody of the SONS OF LIBERTY in this Town, a moſt *deteſtable* Object, lately tranſported to *America*—Notice being given, a Jury was ſummon'd, and a Bill found, ſetting forth that the ſaid Priſoner did on the firſt of November laſt, endeavour to make its Appearance in a forcible Manner, and in Defiance of the known and eſtabliſh'd Laws of the Britiſh Conſtitution, to deprive the Subject of his Rights and Privileges, &c.

Figure 4. Boston Gazette, February 24, 1766. American Antiquarian Society

newspaper published a coffin with the epitaph, "The last Remains of The PENNSYLVANIA JOURNAL, which departed this Life, the 31st of October, 1765. Of a STAMP in her Vitals, Aged 23 Years." These images, along with essays describing funeral processions for "Liberty" and the colonies' freedom, tapped into a mode of emotionally saturated protest that had long used mourning as a means of expressing colonial discontent.[98] Other newspapers around the colonies followed suit in the last week of October.

In areas where British officials were able to enforce the Stamp Act, printers had to resort to subterfuge to publish similar protests. Isaiah Thomas, who was working in Halifax at the time, recounted later in life how he read the October 31 issue of the *Pennsylvania Journal* and wanted to reuse many of the same devices. Because he had already gotten in trouble with colonial authorities for printing pieces opposed to the Stamp Act, Thomas felt uncomfortable simply reproducing Bradford's language and format in the *Halifax Gazette*. Instead, he contrived "an expedient . . . to obviate that difficulty" by writing an article that shifted blame away from the printer-editor. Thomas suggested instead that "a number of [his] readers" had requested "a description of the extraordinary appearance of the *Pennsylvania Journal* of the 30th of October

Thurſday, *October* 31, 1765. THE NUMB. 1195.

PENNSYLVANIA JOURNAL; AND WEEKLY ADVERTISER.

EXPIRING: In Hopes of a Reſurrection to LIFE again.

I AM ſorry to be obliged to acquaint my Readers, that as The STAMP-ACT, is fear'd to be obligatory upon us after the *Firſt of November* enſuing, (the *fatal To-morrow*) the Publiſher of this Paper unable to bear the Burthen, has thought it expedient to STOP a while, in order to deliberate, whether any Methods can be found to elude the Chains forged for us, and eſcape the inſupportable Slavery; which it is hoped, from the juſt Repreſentations now made againſt that Act, may be effected. Mean while, I muſt earneſtly Requeſt every Individual of my Subſcribers, many of whom have been long behind Hand, that they would immediately Diſcharge their reſpective Arrears, that I may be able, not only to ſupport myſelf during the Interval, but be better prepared to proceed again with this Paper, whenever an opening for that Purpoſe appears, which I hope will be ſoon. WILLIAM BRADFORD.

Remember, O my friends! the Laws, the Rights,
The generous plan of power deliver'd down,
From age to age, by your renown'd forefathers;
O let it never periſh in your hands!
But piouſly tranſmit it to your children.
Do thou, great Liberty, inſpire our ſouls,
And make our lives in thy poſſeſſion happy;
Or our deaths glorious in thy defence. ADDISON's Cato.

LIBERTY is one of the greateſt Bleſſings, which human beings can poſſibly enjoy: When we are deprived of this earthly happineſs, we are fettered with the Chains of inimical ſervitude. Nations, who are born for the mutual ſupport of each other, ſhould preſerve a ſteady attachment to the welfare and happineſs of that nation with whom they are united, that their mutual alliance of friendſhip might be ſincere and permanent. When this union is ſeparated by the illegal encroachments on that Liberty, which is the Soul of Commerce, and the Support of Life, it degenerates into implacable Enmity, which in time grows inveterate, and finally recoils upon thoſe who have been the means of its unhappy diſſolution. The *Liberty of the Preſs* has very juſtly been eſteemed one of the main Pillars of the Liberty of the People. While this is maintained, the firſt Steps to Oppreſſion are detected, and the Attention of the People ſeaſonably awakened. When this is ſuppreſſed, the Suſpicion of the People, and their Ruin may admit of ſo ſudden a tranſition, as renders the Succeſs of the firſt impracticable, and the Miſeries attending the latter unavoidable. So dangerous is this to lawleſs Power, that the fartheſt approaches to it are reſolutely oppoſed, or rigorouſly puniſhed. So eſſential is this to Freedom, Property, and Happineſs, that the moſt plauſible Attempts to curtail it even in the ſmalleſt Degree, have always been moſt ſtrenuouſly oppoſed by the virtuous, free, and unbiaſſed Patriot. It is the Priviledge of Britons to ſpeak Truth with Impunity, and even to fear no Danger from ſpeculative Error whether in Religion or Politicks. The want of attending to this has produced needleſs Enquiries, and unjuſtifiable Cenſures of what is true in Fact or no more than falſe in Speculation.--But how unhappily is it to be debarr'd from this laſt Relief in a dangerous and ſickly State! How melancholly to pine and die while no kind Phyſician is allowed to explore the Cauſe, or preſcribe the Cure of our manifold Diſorders! The love of Eaſe, and abſence of Pain in ſome Diſtempers, is a fatal Symptom of the deſperate Circumſtances of the Patient.

And in all political Diſorders the more contented we are under them, ſo much the worſe are they, and ſo much the worſe are we for them. It is a very happy Circumſtance attending public Virtue and public Spirit, that the more it is vilified, the more illuſtrious it always appears. No Falſhood formed againſt it can proſper, for it at once detects and confutes the darkeſt and moſt inveterate Calumny. But although public Virtue cannot be affected by the Indulgence of the moſt unlimited Freedom of ſpeaking or writing, yet Oppreſſion and Tyranny, as it derives all its Influence from its Secrecy, may be extremely benefited by the Reverſe. For this reaſon, in Countries ſubjected to the inſatiable Demands of Power and Avarice, the firſt Attempts to inſpire People with a juſt Senſe of their Condition, are commonly nipt in the Bud. It is of the laſt Importance to the Views of deſigning Men to ſhut up the moſt ſucceſsful and univerſal Channel of Information from the People, when they are forming ſuch Schemes as need only to be known in order to be Oppoſed. Beſides the Deprivation of our whole Liberty may be juſtified on the ſame *Principles* as the Deprivation of any individual Part, ſuch as the Liberty of the Preſs undoubtedly is.

How amiable is the Enjoyment of Liberty! But how deteſtable are the Bonds of Servitude! 'Tis therefore ſincerely to be hoped, that the old *New-England* Spirit ſo exemplarily free in former Times, will never condeſcend in Submiſſion to new and unwarrantable Reſtrictions.

A Day, an Hour of virtuous Liberty,
Is worth a whole Eternity in Bondage.

May we all as loyal Subjects, and free born Britons exert our utmoſt to preſerve the Rights and Liberties of our Country, in a Manner that ſhall add Honour to our Endeavours; that future Poſterity may reap the Benefit, and bleſs the Hands which were the Inſtruments of procuring it.---

That Glory then, the brighteſt Crown of Praiſe,
Which every Lover of his Country's Wealth,
And every Patron of Mankind deſerves;
Will gracefully adorn ſuch Patriot's Deeds,
And leave behind an Honour that will laſt
With Praiſe immortal to the End of Time.

Thurſday laſt arrived here the ſhip Philadelphia Packet, Capt. Budden, from London, by whom we have the following advices.

R O M E, *July* 24.

THE harveſt in this country hath not proved ſo good as we hoped. This event hath engaged the congregation eſtabliſhed for inſpecting into the ſupplies of proviſions for this capital, to ſeek all poſſible means to prevent a freſh ſcarcity.

St. James's, Auguſt 17. The king has been pleaſed to appoint the moſt honourable the Marquis of Rockingham to be lord lieutenant of the weſt-ridings of the county of York, and of the city of York, and county of the ſame city; and alſo Cuſtos Rotulorum of the north and weſt-ridings in the ſaid county of York and of the city of York, and county of the ſame city; and Ainſty, otherwiſe Aynſtry, of York.

The king has been pleaſed to appoint the right hon. William Earl of Dartmouth, Soam Jenyns, Edward Eliot, John York, George Rice, John Roberts, Jeremiah Dyſon, and William Fitzherbert, Eſqrs; to be commiſſioners of trade, and for inſpecting and improving his majeſty's plantations in America, and elſewhere.

The king has been pleaſed to grant unto the right hon. Richard Viſcount How, the office of treaſurer of his Majeſty's navy.

St. James's, Auguſt 17. By the laſt letters from Col. Deſmarets, his majeſty's commiſſary at Dunkirk, we are aſſured, that orders were given by the French miniſtry, for immediately ſetting about the demolition of the Jettees, which are the ſupport of the harbour of Dunkirk.

Warſaw, Auguſt 1. The tribunal of Great Poland, held at Poſnania, has granted permiſſion to the Lutherans at Loböcutz to open their church, which has been ſhut near twenty years, to provide a miniſter, and to perform divine ſervice in public.

Corunna, July 17. The detachment of one hundred men draſted for Louiſiana, are on the march for Ferrol, where they are to embark on board the Unicorn frigate with a governor, two capuchin friars, a commiſſionary at war, and ſome civil officers.

Cadiz, July 23. Letters brought by the laſt poſt from Gibralter ſay, the report before ſpread, that the Algerines have killed their Dey, and declared war againſt all the European powers except England and France, proves not true.

L O N D O N.

Auguſt 17. On Thurſday at the king's arms tavern in Cornhill, an elegant entertainment was given by the committee of North-American merchants to Richard Glover, and Charles Garth, Eſqrs; when thoſe gentlemen received the thanks of that body, for their endeavours to prevent the ſoldiery from being billeted upon the private houſes of their fellow-ſubjects in America.

Part of a letter from an officer in the Eaſt-India ſervice, dated from the Arietur camp, January 8, 1765.

"In my laſt I acquainted you that we did at laſt reduce Madure. The army has ſince conquered the Arietur country for the Nabob, of 100,000l. revenue a year. We are now under orders to attack another chief, or polygor contiguous to this country; both chiefs have mutually maintained an independency of the Nabob till now; meerly on account of the impenetrable woods they are poſſeſſed of. You certainly have heard before of the memorable battle Major Munro gained at Bengal over Suja Dowla, one of the moſt formidable powers of India. The conſequence of this battle gives the company the command of trade in the greateſt part of the Mogul's dominions; and, without exaggeration, the Eaſt-India company at preſent may be brought in compariſon with Alexander the Great, whoſe command, from the river Indus to the river Ganges, was not ſo much reſpected as their's."

It is ſaid the new m-----y, taking into conſideration the preſent deplorable ſituation of the Canadians, have determined to take up all the Canada bills at par, with intereſt to the preſent time; and afterwards to demand, in the moſt ſpirited terms, *immediate* and *full* payment of France, under pain of all the conſequences that can reſult from a refuſal.

The new lords of trade and plantations will hold a board on Monday next, for the firſt time, at the Cockpit Whitehall.

We hear the rent rolls of the ſeveral proprietary eſtates in America, obtained by former grants under the crown are ordered to be made out, as alſo an eſtimate of the annual produce of their land tax, in order to introduce a more equitable form of levying his majeſty's revenues in that part of the world.

They write from Gibralter, that Engliſh officers and ſeamen are engaging both there and at Minorca by foreign agents, to ſerve on board his Sardinian majeſty's ſhips of war in the Mediterranean.

Auguſt 20. The right honourable the Earl Cornwallis, lieutenant colonel to the 33 regiment of foot, is appointed one of his majeſty's aids de camp, with the rank of colonel of foot in the army.

We are informed, that a gentleman lately very popular in this country, is ſoon to reſide at Louſanne, in Switzerland, where he intends publiſhing his friend Mr. Churchill's poems, with explanatory notes; and we are likewiſe informed, that he has an intention of publiſhing, at the ſame place, a hiſtory of England wrote by himſelf.

By a veſſel arrived at Guernſey from Belleiſle there is advice, that the French are erecting ſeveral batteries at Sandy Bay, on the ſouth eaſt of the iſland, and in the great road where the deſcent was made in the late war by the ſeamen and troops under Admiral Keppel and General Hodgſon.

Aug. 21. We hear lord Viſcount Spencer is ſhortly to be created an earl.

All thoughts of any farther changes are ſaid to be entirely laid aſide.

It is reported, that a perſon of high rank, on being lately offered a great employment, refuſed it, ſaying, "that he could not poſſibly accept of it, conſiſtently with the love he bore to the Britiſh nation, which would ever be the object of his care and attention."

Monday ſome diſpatches ſaid to be of importance, were received here from Holland; but the ſubject has not yet tranſpired.

Private letters from Paris mention, that the true reaſon [illegible]

An Emblem of the Effects of the STAMP. O! the fatal STAMP.

Adieu, Adieu to the LIBERTY of the PRESS.

Figure 5. Pennsylvania Journal, October 31, 1765. American Antiquarian Society

last, 1765." He could comply, he argued, only by reproducing the text and imagery from that newspaper.[99] Thomas thus used one of the conventions of newspaper publication to his own advantage. Readers often wrote to printers, asking them to publish particular excerpts, letters, or pieces from other newspapers or magazines, and printers often obliged, citing popular demand. Here, the rhetorically nimble Thomas, himself a militant protester, created space for himself rhetorically to publish protest imagery and text under the pretext of responding to a reader's request.

As 1766 dawned, some aspects of colonial commercial life remained shuttered. Merchants continued with great success the nonimportation agreements designed to influence London merchants to lobby on their behalf. And lawyers' inactivity kept courts shuttered. Criminal cases were in most cases unaffected, but nearly all civil cases were delayed as lawyers fretted about the validity of paperwork filed or judgments entered by judges without stamps as long as the law remained technically in effect.[100] Still, colonial life had returned to normal by the winter of 1766. This was largely true of the printing trade too—with the exception of the few printers who refused to publish their newspapers until the act was repealed. By that time, it looked as if they would get their way: news from England indicated that Parliament planned to revise or repeal the Stamp Act. In February Franklin appeared before the House of Commons to defend the Americans' arguments for repeal, and William Strahan forwarded the clerk's transcript to David Hall in April so he could be the first to offer it for sale as a pamphlet.[101] When the news finally arrived from England in May that Parliament had repealed the act, printers reacted with joy, having avoided any penalties for their nominally illegal printing. They printed news of the repeal with encomia to Henry Conway, secretary of state for the colonies, Colonel Isaac Barré, and William Pitt, the three men colonists saw as most responsible for defending their rights and bringing about the repeal. On the other hand, printers throughout the colonies also published numerous letters from "friends of America" in England, warning them that overly gleeful responses to news of the repeal would not be greeted warmly in the mother country. In many places, reactions were accordingly muted.

The return of business was not entirely as usual, as the fissures created by the Stamp Act remained. Printers were slow to forget the Stamp Act crisis; in fact, several almanacs in 1767, and even beyond, made special note of March 18, the anniversary of the repeal, in their calendars.[102] The crisis also began to redefine the role of printers in the political life of the colonies. As editors and publishers who managed the circulation of political news, they proved vital to the opposition to and nullification of the Stamp Act. In doing so, they attempted to negotiate and reconcile their commercial interests with the political concerns that guided them and their readers. Finally, the crisis created alliances and channels of communication that began to align printers with political leaders, particularly among Patriots. These connections would reap dividends in the years to come.

The Business of Protest

Printing against Empire

After the end of the Stamp Act crisis in 1766, the focus of imperial policy turned away from taxing printed matter to other concerns. For a time, colonists felt they had won a total victory. Over the next several years, however, that belief would prove ephemeral as Parliament passed various acts that reinforced its declaration that it could legislate for the colonies "in all cases whatsoever." Taxes on consumer goods, regulations about supplying troops quartered in North America, the very presence of troops in colonial port towns, and persistent attempts to reduce the power of the colonial assemblies all sparked protests among colonists. Though none of these parliamentary actions directly targeted printing as a business, printers mobilized to strengthen ties they had developed in 1765, to publish more protest literature, and to solidify the pathways they used to improve the quality of the news they received. By the time Parliament passed the Tea Act in 1773, printers had formalized a set of connections with protest leaders and extralegal groups that circulated political news throughout the colonies in newspapers, pamphlets, and other forms of print.

The period between the Stamp Act in 1765 and the Tea Act of 1773 was a time of increasingly polarized debate about imperial policies in the colonies. Whereas in the mid-1760s most printers placed their commercial self-interest at the center of their decision making, by the early 1770s many used politics as a barometer on at least an equal basis when considering how to deal with events related to the imperial crisis. To enhance their networks and communications pathways during these years of protest and debate, printers worked to expand their connections with political and commercial leaders within their communities and to extend their networks of contacts with other printers throughout North America (and, when possible, Europe). To do so, printers worked to enrich and thicken the media in which they worked.

Printers always sat at the intersection of oral, written, and printed forms of debate, but during the years after the Stamp Act crisis those forms came to serve as a means of embodying the networks that printers were creating. By replicating conversations, letters and other manuscript sources, and excerpts from newspapers and pamphlets, printers made their newspapers and other publications the central hub for understanding debates about the imperial crisis.

Printers, for example, published the letters of the Pennsylvania Farmer in nearly every colonial American newspaper and in several pamphlet editions in the winter and spring of 1768 at the height of protests against the Townshend Acts. In doing so, printers strengthened intercolonial channels of communication in part by reprinting and distributing this popular set of essays for commercial gain. At the local level, Boston printers worked with the Boston Committee of Correspondence as it began to integrate the town's hinterland into a single regional network for circulating Patriot political communications. The committee's efforts to expand were able to build on the already well-developed commercial connections of printers, but printers also pushed back to protect their commercial interests when necessary. Both examples demonstrate the importance of the printer's skill as an editor and compiler in shaping the flows of political information that arrived in his office. Likewise, the protests against tea in the fall and winter of 1773 took advantage of the networks and pathways that printers had worked on for years in order to effectively rebut the tax and the delivery of the tea.

Such case studies shed light on the range of media that printers employed to communicate effectively in local and regional contexts. During the late 1760s and early 1770s, the printing trade thickened its relationships both within the trade and among printers, political and commercial leaders, local governments, and extralegal organizations such as the Sons of Liberty. The protests against the Townshend Acts spurred further cooperation among these groups, which continued to develop both during crisis moments and through more mundane everyday interactions. Printers relied on modes of transmission such as the post office to circulate news and information on a broad geographic scale. Locally, the key mode of transmission for political news was oral communication and personal contact with the printer. To facilitate these conversations, printers placed their offices as close to the nerve centers of local commercial and political life as they could. Most were near taverns or coffeehouses, the post office, a market or exchange, or town wharves. Such locations facilitated the easy handoff of essays by local luminaries, such as the "Humphry Ploughjogger" letters written by John Adams. Printers also made their newspapers and pamphlets a printed extension of political debate that occurred in public places such as taverns and coffeehouses. As one historian of public houses has noted, they "were wellsprings of indigenous textual productions in Philadelphia, and many pamphlets and newspaper features mirrored tavern speech precisely in order to sway a readership that continued to hold oral discourse in high regard."[1] In addition to reflecting local political debates, newspapers thus served as an outlet for imperial, colonial, and municipal agents to announce legislation, publish proclamations, and update the public on various governmental doings.

The circulation of news locally thus depended largely on word of mouth and hand-to-hand distribution of newspapers. Local subscribers usually picked up their copies of the newspaper at the printing office, but some printers also used news carriers, either their own apprentices or servants for hire, to deliver papers. Hugh Gaine, for example, paid his news carriers about three shillings per year per customer to deliver his *New-York Gazette, and Weekly Mercury*.[2] Printers also used almanacs as a means of distributing political news locally. Nathanael Low, in a preface to his *Astronomical Diary* for 1771, wrote, in reference to the Townshend boycotts and "wear homespun" campaigns: "The Cause we are engaged in is just;—the Prize we are contending for, is inestimable.—Let us therefore act with Resolution and Fortitude: This will overcome the greatest, and otherwise most insuperable Difficulties—With Strength and Vigor let us maintain a political Warfare."[3] The language was standard for protest literature, but including such arguments in almanacs, the cheapest and most popular print genre, helped advance them with people of all classes. In that way almanacs usefully supplemented newspapers as a way to disseminate widely political news and opinion.

Though colonial printers had earlier eschewed open political controversy, by the 1760s the political opportunities available to printers broadened, especially in the major port towns where printers had established multiple newspapers. In these locations, a competitive market sustained a wider spectrum of debate independent of the issues related to the relationship between empire and colonies. Printers in these towns—Boston, New York, Philadelphia, Williamsburg, and Charleston—began to associate increasingly with one side or the other in local and imperial political disputes. These associations continued after the Stamp Act, when several printers became involved in the groups of the Sons of Liberty that sprang up in the colonies. In Philadelphia, William Goddard arrived in 1767 in large part because the antiproprietary party led by Joseph Galloway (and behind the scenes by Franklin) found David Hall to be an unsuitable publishing partner.[4] In New York, printer John Holt relied on the support of the Livingston family faction for his commercial survival.[5] Some printers also included outsiders in the preparation of their publications. John Adams noted in his diary in 1769 that he spent an evening at the printing office of Edes and Gill "preparing for the Next Days Newspaper . . . Cooking up Paragraphs, Articles, Occurences, &c." with Gill present, along with noted political activists Samuel Adams and James Otis.[6] In creating such connections, printers could ensure the stability of their businesses through steady work and rely on a steady flow of material for their publications.

By the early 1770s, Patriot leaders extended these connections across the colonies through the establishment of committees of correspondence. Following a standard function of colonial legislatures, these groups operated with the sanction of either those bodies or their local towns. In the latter case, they were, by definition, extralegal, having no sanction under most colonial charters or by imperial authority. Because of their close ties through preexisting information pathways, printers frequently joined these committees or served as key nodes for the transmission of their news, including Edes and Gill and Isaiah Thomas in Boston, John Holt in New York, and Peter Timothy in Charleston. These formal connections for printers cemented work they had undertaken for nearly a decade to build their networks. Printers who managed these relationships well positioned themselves to take advantage of an increasingly partisan atmosphere for political publishing.

Protesting the Townshend Acts in Published Letters

In the wake of the Stamp Act crisis settlement, imperial politics briefly quieted as American colonists celebrated what they thought was a clear victory. Meanwhile, officials in London continued to seek ways to retire debts and fund the expenses of empire, with the additional motivation of implementing the Declaratory Act to demonstrate that Parliament did indeed have the power to legislate over the colonies "in all cases whatsoever." That opportunity came in 1767. Confident that the dust had settled from the Stamp Act dispute, Charles Townshend, the chancellor of the exchequer, proposed to assert parliamentary authority to tax American colonists through a series of duties placed on various consumer goods, including tea, paper, china, paint, lead, and glass.[7] Colonists who thought the Stamp Act repeal in 1766 had settled matters felt betrayed and immediately moved to rekindle their protests. Resistance leaders engaged the merchants of major towns in nonimportation agreements, a strategy that had seemed effective in marshaling support in Britain in the previous dispute.[8] Print media abounded with arguments against (and a few in favor of) the new duties, and their circulation in newspapers and pamphlets throughout the colonies reflected a growing effort to coordinate publication and consciously shape their distribution.

Two genres proved critical to these efforts: pamphlets and published letters. Pamphlets were as crucial to printers' livelihoods and the anti-imperial resistance movement as newspapers. The arguments laid out in the pamphlet literature of the imperial crisis, in fact, established much of the ideological groundwork for the Revolution.[9] Printers found pamphlets useful sales tools as well. They could be published quickly and sold relatively cheaply. That speed meant that pamphlets could serve as a medium of rapid response that allowed for greater elaboration than the newspaper: a sermon that struck a chord with the congregation, an oration that particularly moved the crowd in the public square, or thoughts on the latest parliamentary act within the space of a few days could become a forty- or fifty-page octavo pamphlet, stitched and sold for just a few pennies. In 1773 the anniversary oration for the Boston Massacre by Benjamin Church went through three editions in less than three weeks after it was first delivered before a crowd in Boston.[10] Because printers exchanged pamphlets as well as newspapers and books to sell in one another's printing offices and adjoining bookshops, they also circulated rather easily in the colonies.

Published letters appearing in either newspapers or pamphlets formed a second key genre for political argumentation. This form allowed writers to engage with a well-defined audience (either a specific individual or a more general group of people) and encouraged readers to feel they were being included in a private conversation. Newspaper printers mined correspondence with local noteworthies to produce news for publication and invited authors to submit essays framed as letters.[11]

John Dickinson took advantage of both the newspaper essay and pamphlet as forms to advance his cause during the imperial crisis. Born into a prominent Maryland family and raised for much of his childhood in Delaware, he trained in the law at the Middle Temple in London. When he returned to Delaware, he built a lucrative practice and became active in the colony's assembly. In the early 1760s he moved to Philadelphia and became a prominent voice in Pennsylvania politics during that decade. As a fellow traveler of the Quakers, he had been among the most vocal in defending the Quaker constitution against the movement for royal government led by Benjamin Franklin and Joseph Galloway.[12] He was among the de facto leaders of the Stamp Act Congress in October 1765 in New York, where he drafted the Resolves of the Congress and commenced a friendship with leaders from other colonies, including Samuel Adams and James Otis of Boston. He also engaged in a pamphlet exchange with the Barbados Assembly, excoriating it for allowing the Stamp Act to take effect on the island.[13] His most famous moment would come two years later as colonists debated how to respond to the Townshend Acts.

Dickinson outlined the most prominent of the arguments against the Townshend Acts in the letters of "A Farmer," more generally known today as the "Pennsylvania Farmer," a pseudonym he used to evoke pastoral simplicity. Twelve in number, the letters first appeared in Philadelphia's *Pennsylvania Chronicle* and *Pennsylvania Gazette* between December 1767 and February 1768. In them, Dickinson carefully laid out his argument that the American colonies were gravely threatened by parliamentary oppression. Dickinson directed most of his energy in the letters to attacking the Townshend Acts but also devoted significant space to protesting Parliament's censure of the New York Assembly, which had refused to obey a quartering act that required colonies to provide supplies to British soldiers stationed within their borders. Protesting these laws—which he believed violated the constitutional bonds between Parliament and the American colonists—was for him a matter of conscience and a vital act of civil disobedience. Repealing the duties, he argued, was critical to restoring balance and unity to the British constitution.[14]

Dickinson mobilized standard Whig conceptions of liberty and freedom to claim that the duties were unconstitutional for several reasons.[15] First, he allowed that Parliament had the right to regulate the trade of the empire but argued that Parliament could not justify raising a regular revenue from those regulations. In this respect, he noted, the new duties on goods were no different from the requirements of the Stamp Act, which colonists had successfully resisted just two years prior. He also parried several arguments of the acts' proponents. In Letter VII he countered the view that the duties were too small to be of consequence, suggesting instead that it was "the very circumstance most alarming to me" because Parliament "intended by *it* to establish a *precedent* for future use."[16] He also debunked the claim that the taxes were to provide security for the colonies themselves by noting that the established colonies had paid for their own security long before Parliament ever became interested in them and that Parliament was not asking Canada or Florida to contribute to their own defense.[17]

He also invoked key features of the classical republican tradition in America, arguing that losing control over taxation would lead to corruption, the convergence of power in illegitimate places, and the oppression of standing armies.[18] To buttress these claims, Dickinson referred to a wide variety of precedents from the Stamp Act to ancient Rome and the Glorious Revolution.[19] Dickinson had very little to offer, however, in terms of protest mechanisms—leaving that to the printers who made hay with his argument. He urged a unified response and, at the conclusion of the final letter (Letter XII), used the example of the Stamp Act protests as a model of working for repeal of the Townshend duties: "Is there not the strongest probability, that if the universal sense of these colonies is immediately expressed by RESOLVES of the assemblies, in support of their rights, by INSTRUCTIONS to their agents on the subject, and by PETITIONS to the crown and parliament for redress, these measures will have the same success now, that they had in the time of the *Stamp-act*?"[20] The letters crystallized for many colonists the disagreements with British imperial policy and took North American print media by storm.

The Pennsylvania Farmer letters were among the most successful serialized newspaper essays and political pamphlets in the late colonial era. They achieved a greater circulation than any other piece of propaganda until *Common Sense* in 1776. Printers took what appeared on the page to be locally written letters for Philadelphia newspapers and made them an intercolonial sensation—just as Dickinson had intended. He seemed to have written all the letters before publication began in December 1767 and to have given drafts to both David Hall, publisher of the *Pennsylvania Gazette*, and William Goddard, publisher of the *Pennsylvania Chronicle*. Dickinson then polished them as they appeared in ensuing weeks but gave the updates only to Hall. His doing so suggests that his relationship with Hall was probably somewhat stronger than with Goddard (who regularly irked those with whom he worked). It also meant that the versions in the two newspapers differed, even though both claimed that theirs were authoritative. Dickinson noted the discrepancy in a letter to James Otis of Boston: "The only correct one publish'd here, is printed in the *Pennsylvania Gazette* of Hall and Sellers. I find that the 'Letters' publish'd to the Eastward, are taken from our *Chronicle*, which being incorrect, I should be glad if you would be so kind as to mention to any of the Printers you may happen to see, that the *Gazette* is much the most exact."[21]

After their initial publication in Philadelphia, one or more of the twelve letters appeared in nearly every newspaper in North America (table 1). These

TABLE 1. *Publication of the "Pennsylvania Farmer" Letters in American Newspapers, 1767–1768*

Newspaper	Town	Publication Date of Letter I	Average Time Lag (in days)
Pennsylvania Gazette	Philadelphia	Dec. 3, 1767	3
New-York Mercury	New York	Dec. 7, 1767	12
Pennsylvania Journal	Philadelphia	Dec. 10, 1767	3
New-York Gazette; or, the Weekly Post-Boy	New York	Dec. 10, 1767	11
New-York Journal	New York	Dec. 10, 1767	10
Maryland Gazette	Annapolis	Dec. 17, 1767	23
Providence Gazette	Providence	Dec. 19, 1767	24
Connecticut Courant	Hartford	Dec. 21, 1767	39
Boston Gazette[a]	Boston	Dec. 21, 1767	18
Boston Evening-Post[b]	Boston	Dec. 21, 1767	—
Boston Chronicle[c]	Boston	unknown	22
Virginia Gazette[d]	Williamsburg (William Rind)	Dec. 24, 1767	42
Connecticut Gazette	New London	Dec. 25, 1767	31
Newport Mercury	Newport	Dec. 28, 1767	25
South-Carolina Gazette	Charleston (Timothy)	Jan. 4, 1768	39
South-Carolina Gazette; and Country Journal	Charleston (Crouch)	Jan. 5, 1768	40
Virginia Gazette	Williamsburg (Purdie & Dixon)	Jan. 7, 1768	40
New-Hampshire Gazette	Portsmouth	Jan. 8, 1768	70
Georgia Gazette	Savannah	Jan. 27, 1768	64

Note: The average time lag measures the delay between the date each letter was published in Philadelphia and the date it was published in the newspaper listed. The table represents nearly all newspapers for 1767 and 1768. Publication for the *New-York Gazette* (published by William Weyman) and the *Germantowner Zeitung*, could not be determined; no extant copies of the *Germantowner Zeitung* exist for the relevant period (see Edward Connery Lathem, *Chronological Tables of American Newspapers, 1690–1820: Being a Tabular Guide to Holdings of Newspapers Published in America through the Year 1820* (Barre, MA: American Antiquarian Society, 1972). Whether any Caribbean or Canadian newspapers reprinted the letters is unknown as sufficient copies do not exist for the time period.

[a] Average is based on ten of the twelve letters.
[b] No average was calculated because only four letters were found.
[c] Average is based on nine letters.
[d] Average is based on seven letters.

letters traveled ordinary paths for newspapers, which meant that they were published within a few days in New York, about one to two weeks later in Boston, and at varying times for other newspapers in New England and the mid-Atlantic. Lag times to the South were often longer because of gaps in the postal service. The letters, for example, took about two months on average to reach Savannah, where they appeared in James Johnston's *Georgia Gazette*.[22] Several newspapers appear not to have published the letters at all, including two newspapers published by Boston loyalists: the *Boston News-Letter*, published by Richard Draper, and the *Boston Post-Boy*, published by John Green and Joseph Russell. A third newspaper, the German-language *Wöchentliche*

Pennsylvanische Staatsbote in Philadelphia, was published by John Henry Miller, who was not a loyalist; he may simply have decided not to translate the pieces into German. In New Haven, the *Connecticut Journal*, published by another branch of the Green family—the brothers Samuel and Thomas Green—published only the first of the twelve letters. Otherwise they appeared in nearly all American newspapers between December 1767 and April 1768.

Most of the newspapers that published the letters seem to have done so as soon as they arrived in town, but one newspaper stands out for its slow response. The *New-Hampshire Gazette*, published by Daniel and Robert Luist Fowle, typically published news appearing in Boston papers with only a few days' time lag—Portsmouth was, after all, just sixty-five miles away. Yet the Pennsylvania Farmer letters appeared in the *New-Hampshire Gazette* long after they had been published in Boston—nearly two months later, on average. In fact, it took less time for the letters to appear in the *Virginia Gazette* in Williamsburg or to reach Savannah.[23] Similarly, yet another Green family printer, Thomas Green of Hartford, also delayed printing the letters. The *Connecticut Courant* published the letters on average nearly six weeks after their first appearance in Philadelphia, long after they had already appeared in other New England newspapers. Neither Fowle nor Green offered a reason for the delay. Perhaps they opposed the argument of the essays, or they may simply have assumed their readers would have already seen them. Regardless, their eventual decision to include the letters in their newspapers indicates broad interest among the reading public.

Printers took advantage of the letters' popularity to publish them in additional formats. Eleven pamphlet editions of the letters appeared between 1768 and 1774, including two editions in Boston, four in Philadelphia, and one each in New York and Williamsburg.[24] The printers who produced these editions seem to meet one of three criteria: They were Philadelphia printers, where the letters originated (Hall and Sellers and William Bradford both produced multiple editions); they had strong ties to the resistance movement (Bradford, John Holt, and Benjamin Edes were all members of the Sons of Liberty); or they represented the larger and more successful printing offices in the colonies, whose size allowed them to undertake additional work at a rapid-fire pace and thus take advantage of sudden opportunities to publish.

The letters generated a flurry of responses, ensuring that discussion about them would circulate for months after their initial publication. Encomia to the Farmer proliferated in printed letters that appeared in numerous newspapers through the process of reprinting. Town meetings across the colonies thanked Dickinson for his contribution to the American cause. For example, the town of Boston met in late March and, as reported first in the *Boston News-Letter* on March 24, 1768, declared that "'Tis to YOU, worthy SIR! that AMERICA is obliged, for a most seasonable, sensible, loyal and vigorous Vindication of her invaded Rights and Liberties." They asked that he allow them "to intrude upon your Retirement, and salute The FARMER, as the FRIEND of AMERICANS, and the common Benefactor of Mankind."[25] The Farmer also received plaudits in newspapers from such diverse places as Providence, Rhode Island, in June, and Cecil County in the northeastern corner of Maryland in August.[26] The publication of the letters, nurtured by printers, galvanized support for the tactic of nonimportation at a moment of crisis.

A few newspapers, most notably the *Pennsylvania Chronicle*, carried counterarguments to the Farmer. The *Chronicle*'s owners were the staunch anti-imperialist William Goddard and the loyalist Joseph Galloway. According to Goddard, Galloway forced him to print the attacks on the Farmer.[27] These essays, signed with various pseudonyms, ridiculed the Farmer's arguments. But this counternarrative lacked the force of constant circulation. The essays remained so popular that published rebuttals found little traction in other colonial newspapers.

The letters also received notable attention across the Atlantic. Benjamin Franklin supplied a preface to a London edition published in 1768 by John Almon (Franklin also provided the letters themselves to the London printer).[28] Franklin noted, "The Author is a gentleman of repute in that country for his knowledge of its affairs, and, it is said, speaks the *general sentiments* of the inhabitants."[29] The collected letters thus allowed Franklin, as an imperial representative, to demonstrate colonial opinion at its most eloquent. Almon's edition, with Franklin's preface, was reprinted in both Dublin and Paris.[30] The London edition was reviewed in the *Gentleman's Magazine*, which attempted to rebut Dickinson's claims. In response to his argument that the colonists required representation in Parliament to be legitimately subject to taxation, "it may be here observed, that admitting the principle, the British parliament cannot legally tax the inhabitants of London."[31] George Grenville even referred to the letters in a debate in Parliament about how to respond to the colonial boycotts. According to William Strahan, Grenville gave a lengthy speech during which he "had in his Pocket the Farmer's Letters, which he frequently quoted, and called them *libellous throughout*."[32] Dickinson's arguments thus reached the highest levels of the imperial administration and were taken as a worrisome indicator of American opinion.

Printers circulated Dickinson's essays throughout North America and imperial London during the winter and spring of 1768. Their efforts operated through overlapping circuits of oral, manuscript, and print communication, as the publication of the letters sparked town meetings, correspondence, and printed replies. The Letters crystallized in clear language an assortment of anti-imperial arguments that had been circulating for several years and galvanized resistance to the Townshend duties, in large measure because Dickinson was reframing mainstream arguments with cautious language and a patriotic tone. Printers encouraged and took advantage of the letters' popularity, ensuring their wide distribution in newspapers and pamphlets, two of the most important print media for generating anti-imperial resistance and its rationales.

The Local Context: Forging Alliances Outside the Trade

From the Stamp Act crisis onward and in particular after the Townshend Acts, printers collaborated more closely with local political and commercial leaders. To facilitate communications and political strategy, printers of all political stripes sought to integrate themselves more thoroughly into governmental and extragovernmental networks. In some cases, printers themselves served in local government, on committees, or (for those opposed to imperial policies) among the Sons of Liberty. Despite the differences in economic and social class between printers and most leaders, printers proved themselves to be far more than pawns of the powerful. They were therefore often not "meer mechanics," notwithstanding the protests they made in their publications. Instead, printers in each of the colonial political centers sought to forge a relationship that was mutually beneficial for their business and political interests and the various interests of the politically connected men with whom they worked. Their success varied according to the particulars of a situation, but across the colonies printers viewed this collaboration as central to their work in printing and publishing.

The interaction of Boston's printers and the city's Committee of Correspondence provides an excellent case study of local political networks and the production of political news. From very early on in the imperial crisis, Boston was far and away the most politicized place in the colonies. This singularly forced printers to take clearly defined positions with respect to imperial politics more quickly than in other major towns. Furthermore, Massachusetts had the best developed print culture of any of the North American colonies, with four newspapers operating in the colony in 1763 and ten by the outbreak of war in 1775. These presses operated in Salem, Newburyport, and Worcester as well as Boston.[33] During that dozen years, Boston developed a collection of strong Patriot newspapers and several assertive Tory, or pro-government, newspapers alongside a few that remained doggedly neutral, even in the face of enormous pressure. Those printers who openly affiliated with one side tended to work in concert with political leaders to shape the content of their printed materials.

The few printers in Boston willing to be viewed publicly as supporters of British policy depended on government appointments from friendly officials to protect them against the vagaries of popular derision and rage. In fact, the printers' political stand made these officials more than willing to put out a hand for the "friends of government." Though never large in number, these printers served as public outlets for embattled imperial officials. This type of favor was particularly important in sharply polarized Boston, where the offices of John Green and Joseph Russell and of John Mein and John Fleeming both remained steadfastly loyalist.[34] After a particularly rocky spell in 1768 and 1769, the two firms became printers to the American Board of Customs stationed in Boston, and the appointment proved quite lucrative: between 1769 and 1775, the board paid Green and Russell £2,400, and Mein and Fleeming £1,500 for their services.[35] Thomas Hutchinson, the royal lieutenant governor of Massachusetts, even wrote to England to argue the case of John Mein (his partnership with Fleeming ended in 1770) and Green and Russell. He wrote that they had been "sufferers and lost almost all their other customers by refusing to comply with the demands of the late seditious leaders and it will discourage others from adhering to Government if they should be rewarded in this manner for their services they have done."[36] Whether the nearly £4,000 in revenues can be directly linked to Hutchinson is unclear, but his advocacy on their behalf was nonetheless helpful.

In at least one case, the efforts of printers to integrate their work in the local context led to diminished circulation of their news through intercolonial commercial networks. The period from 1768 to 1770 was among the most contentious in Boston because of the presence of several regiments of British regulars within the town. Boston printers capitalized on the debates over the soldiers, including the publication of a *Journal of the Times* that appeared as an extraordinary edition over the course of 1768 and 1769.[37] After more than a year of what residents saw as an occupation, tension mounted in Boston for weeks during the winter of 1770, punctuated by the murder of eleven-year-old Christopher Seider on February 22 by a loyalist named Ebenezer Richardson. The soldiers of the 14th and 29th Regiments, which had occupied Boston since 1768, continued to scuffle with local residents, in particular the apprentices, journeymen, and dockworkers who were in the same age cohort as the soldiers. On the evening of March 5, sentries posted outside the Town House (now the Old State House) fired on a group of young men who had been throwing ice and rocks at the soldiers, killing four men instantly, mortally wounding a fifth, and injuring several others.

The Boston Massacre, as it came to be known, stands out among events during the imperial crisis for the surprisingly limited impact its news distribution had. The Sons of Liberty in Boston did not send out dispatches to other towns. Instead, news spread through ordinary channels as it was published in the Boston newspapers. The event occurred on a Monday night, after the publication of the *Boston Gazette*—the leading outlet for the Sons of Liberty. So the first accounts to appear in print came three days later (on Thursday, March 8) from the *Boston News-Letter* and *Boston Chronicle*, two papers of more loyal sympathies.[38] Each published a relatively brief account of "this most shocking Transaction" that followed a basic chronology of events but offered little support for soldiers.[39] Because of the sympathies of the printers of these papers, few others reprinted their account save for newspapers in Newport and New York.[40]

The more iconic account of the massacre appeared a full week after the event, when the *Boston Gazette* published its next Monday issue on March 12, 1770. In that report, printers Edes and Gill devoted more than a full page of the four-page issue to the massacre, which they portrayed as a "melancholy Demonstration of the destructive Consequences of quartering Troops among Citizens in a Time of Peace."[41] They recounted the actions of the soldiers, leaving out many of the details about how Bostonians responded, and continued through several town meetings that occurred in the aftermath as well as accounts of meetings with Lieutenant Governor Thomas Hutchinson. They surrounded their account with black mourning borders along each column of the second and third pages of the issue and published an image of four coffins with the initials of those who had died (the fifth victim, Patrick Carr, survived until March 14).[42] This account received broader circulation and was reprinted in whole or significant part in at least eight newspapers in the weeks following.[43]

Curiously, however, that was the extent of the publicity. Though historians often point to the Massacre as a seminal event in the run-up to the Revolutionary War, the aftermath of the shootings received little coverage around the colonies. The Massacre, though stark in its potential to provide a lesson about British oppression, nonetheless did not fit a neat narrative that Patriot leaders sought. The violence at the heart of the incident (on both sides) ran counter to years of efforts to portray anti-imperial protests as nonviolent. In the weeks after the Massacre, in fact, Patriots—in particular the Sons of Liberty—sought to moderate the narrative of March 5. Rather than presenting what happened as the actions of a violent (and largely working-class) crowd to effect change, leaders suggested instead that British officials had attempted to provoke the violence. The crowd, Patriots insisted, should follow the lead of the Sons of Liberty lest violence recur.[44] Furthermore, Boston's leaders aimed nearly every account of the Massacre at a narrow audience; most publications, including Paul Revere's famous broadside engraving, were intended to be circulated locally.[45]

Within a few years, the town of Boston began to work more explicitly to connect itself to other towns in Massachusetts as well as other major towns along the Atlantic seaboard. To do so, it established a committee of correspondence. Founded as a branch of the Boston town meeting in 1772, its members had been at the vanguard of anti-imperial protest in Massachusetts since the Stamp Act, and it rapidly became the epicenter of activity for anti-imperial activists. At its creation the town meeting charged the committee to organize resistance within Massachusetts against a proposal to pay the salaries of provincial judges from the royal coffers rather than through money approved by the General Court.[46] Initially, its eponymous task was to contact every other town and municipality in Massachusetts, asking them to pass resolutions in support of their cause and to establish their own committees of correspondence in order to keep communication channels open for coordinated political resistance. Intercolonial resistance in 1772 was relatively limited compared to the late 1760s. That changed the following year, when the Boston committee began to coordinate resistance across the colonies to Parliament's attempts to collect additional tax revenue by granting the East India Company a monopoly on tea sales in America. After the Boston Tea Party and the passage of the Boston Port Act in 1774, it was among the first to call for a new continental congress to meet to discuss an intercolonial response to the closure of the port of Boston.

From its inception, the committee used print to extend the reach of its communications beyond town meetings and manuscript correspondence. The committee issued more than forty imprints under its name in addition to countless newspaper articles and extracts. In fact, the committee even included one member who worked as a printer, Joseph Greenleaf. Once a justice of the peace in Plymouth County, he had moved to Boston in 1771 to write for the anti-imperial side. Accused of writing a piece under the pseudonym Mucius Scævola in Isaiah Thomas's *Massachusetts Spy*, the Massachusetts Council stripped him of his appointment. In 1773, he opened his own printing office, and the following year took over publication of the *Royal American Magazine* from Isaiah Thomas.[47] The committee's activities required a close working relationship with printers such as Edes and Gill and Isaiah Thomas. They were close enough, in fact, that Samuel Adams included the men in doggerel verses he wrote about many of the members of the committee. As he did for other committeemen, Adams satirically described Edes and Gill as "Two Foul mouth Printers__ Tools__ with open Jaws; / For Intrest only not their Country's Cause," implying that their hearts might not be in the resistance effort as much as in the profits to be made from it. As for Thomas, Adams wrote, "A Third we'd mention, but he is so base" that he could "only Rank with Childs __ Revere & Chase," other Boston Patriots whom Adams intended to mock by association.[48] The printers for their part used the committee as a source of income, and the committee needed an outlet to broadcast its arguments widely. Printing offices even proved useful meeting locations. Isaiah Thomas, for example, noted "Meetings of the Patriots at . . . my office" in his autobiographical notes.[49]

The committee needed printers' cooperation from the very beginning. One of the committee's first acts was to distribute printed copies of its introductory pamphlet to Massachusetts towns. To publish it, the committee turned to two firms, the brothers Thomas and John Fleet, who had a reputation for neutrality, and Edes and Gill, the most staunchly patriotic printers in Boston. Convening just a few weeks after the town meeting that created it, the committee voted to distribute the pamphlet to "the Selectmen of each Town and District in the Province, and to such other Gentlemen as the Committee of Correspondence shall direct."[50] To further capture the market for anyone who might sympathize with its cause, the committee also ordered that copies of the pamphlet be delivered to the clergy of Boston, who were also key information nodes, presumably in hopes they would offer sermons on the issue and extend the debate to listeners as well as readers.[51]

The pamphlet, *Votes and Proceedings of the freeholders and other inhabitants of the town of Boston, in town meeting assembled, according to law*, contained a three-part treatise that named the members of the committee, outlined the ongoing struggle over judicial salaries, and enunciated colonial rights.[52] The new committee intended the pamphlet to coalesce its anti-imperial arguments in one place and to facilitate communication among Massachusetts towns about the crisis in order to organize coordinated protest. The pamphlet concluded with "A LETTER of Correspondence to the other Towns," in which the Boston committee recounted its discussions with the governor, Thomas Hutchinson, and its decision to go public with the dispute. Committee members sought from the towns "A free Communication of your Sentiments to this Town, of our common Danger."[53] The pamphlet put the matter in stark terms. If the other towns agreed with Boston, they should state so upfront so that they could all jointly defend their rights. If not, Bostonians would be "resigned to our wretched Fate."[54] The committee knew what answer it was likely to get. Its aim was to project a sense of public unity and broadcast public support for its resistance measures by publishing the towns' letters in Boston's newspapers.

Letters quickly arrived in Boston from throughout the colony. The publication of that correspondence generated a rancorous debate over the terms on which they should be printed. The committee and Boston's printers disagreed fundamentally on the status of the reports and the letters from towns, even debating whether the letters constituted "news." The Committee of Correspondence expected to publish the results of its correspondence in the Boston press and assumed that the town's printers would simply offer space in their newspapers as part of their stewardship of the "public prints." It therefore presumed to offer the letters and proceedings of various towns "to the several Printers for a place in their Papers, if they see fit to publish them free of charge."[55] In other words, the committee reasoned that the materials constituted news that would otherwise be printed.

At first, printers accepted the committee's premise that the letters and resolves from Massachusetts towns were newsworthy, printing them dutifully and apparently without compensation. For instance, the committee received one of its first replies from Plymouth, along with the town's manuscript votes and proceedings. It then "offered" the letter "to Messrs. Draper, & M^{r}. Thomas for a place in their several Papers next Thursday."[56] Thomas did publish the letter and resolves of Plymouth without attribution but noted that it was something he was "desired to publish," a standard phrase used to indicate that a third party had requested publication. This language also reinforced the image of the printer as a mere conduit for information rather than one who controlled its flow.[57] Draper published the material as well, without acknowledging the Boston committee as his source, likely because he wished to keep some distance publicly from an avowedly anti-imperial group. Instead he framed it as "taken from the Evening-Post and the Boston Gazette of Monday last."[58] The first few towns to respond were newsworthy, as was the Boston committee's attempts to create a communications network where none had previously existed. Its early success met with justified interest.

Within a few weeks, however, many of Boston's printers began to balk as the committee's requests piled up. By the end of January 1773, the committee had received responses in the form of letters or town resolves from sixty-two municipalities in Massachusetts and had requested that Boston's printers publish most of them.[59] The committee sensed trouble at its January 19 meeting, noting brusquely in its minutes that the newly arrived town proceedings "be published in the News Papers, provided the Printers, will do it free of charge, as they have with respect to the other Towns, where Votes &c. have been printed."[60] Evidently, however, the printers refused. By the following week, after declaring that "the Printers do not publish them with the desired expedition," the committee decided that it would defray "the Expence of Paper, for the Supplements of three public Prints." (A week later the committee pooled funds to purchase beer for its meetings.)[61] The printers themselves, whatever their political affiliation, proved unwilling to devote entire pages of their newspapers to the Boston committee's mission without recompense. In essence, then, even pro-Patriot printers wanted to treat the resolves as advertisements or notices to be paid for by the committee—or perhaps the town meeting, which had granted the committee its authority. The letters took up valuable advertising space that helped them keep their newspapers going. A defunct newspaper did neither the printers nor the Boston committee very much good.

Printers did continue to function as conduits of other kinds of news for the Boston committee. They broadcast a range of political correspondence on various germane topics and typically shielded the committee's identity as the source of the information. For instance, on December 25, 1772, a committee of men from Providence, Rhode Island, wrote to Samuel Adams about the investigation of the burning of the British frigate *Gaspée* in Narragansett Bay.[62] It sought the advice of Adams and "such of your freinds [*sic*] and acquaintance as you may think fit" about how the Rhode Island General Assembly should respond to a letter of Lord Dartmouth, the secretary of state for American affairs, which was transcribed as part of the letter. Adams delivered the letter to Isaiah Thomas, who published it within a week. He stated only that "the following may be depended upon as a genuine extract" of Dartmouth's letter to the governor of Rhode Island.[63] For the average reader, there was no way to determine how Thomas had acquired the letter and thus no way to assign him an active role in choosing to publish it. By phrasing the prefatory comments in the passive voice, Thomas asked his readers to trust his assurances about the letter's authenticity, allowing him to avoid revealing the partisan political pathways by which the letter traveled from the governor's desk in Newport to the pages of a Boston newspaper.

Newspapers also served as an outlet for essays and other texts authored by committee members. In the same December 1772 issue, Thomas published a "letter from the Country," supposedly written by a man who lived in a town "situated at a great distance from Boston" and who had therefore only just received the Boston pamphlet. The letter railed against the provincial government on exactly the terms that the Boston committee was emphasizing—and not coincidentally.[64] The author actually lived in Boston, not in the country, and his name was Samuel Adams.[65] Adams, in fact, was among the most prolific essayists in Boston. Between 1768 and early 1772 alone, he published at least sixty essays in Boston's newspapers, mostly in the *Boston Gazette* of Edes and Gill. Newspapers and printers were clearly an important part of the Boston committee's communications strategy.

The committee and printers worked in interlocking and overlapping ways to bring printers in New England and beyond into affiliations. For printers, the economic motive was strong to seek out new markets as towns and colonies grew. During the eighteenth century, printing in New England had expanded from its seventeenth-century base in Boston to the other three colonies. All four New England colonies had printing offices by the mid-1750s. By 1775 there were printing offices in Boston, Salem, and Newburyport in Massachusetts; New Haven, Hartford, New London, and Norwich in Connecticut; Portsmouth in New Hampshire; and Newport and Providence in Rhode Island. Isaiah Thomas, for one, was consistently on the lookout for ways to grow his business. In undated notes for an autobiography, he recorded that he "set up a press at Newburyport—applications to set up a press in various parts of the Continent—one at Quebec—agree to establish a press at Worcester."[66]

The committee for its part worked to integrate printers in other locales into its network when possible. One way to do that was by forwarding news it received in manuscript from a particular location back to the printer in that place to publish. For instance, the committee ordered that a copy of resolves from the town of Ashford, Connecticut, offering support for Boston during the Port Act crisis, be delivered "to the Printer of the New London Gazette, desiring him To Print the same."[67] Along the same lines, Samuel Adams struck up a correspondence with Charleston printer Peter Timothy in the early 1770s and continued the connection after the Tea Act crisis, when Timothy had become secretary to the Charleston Committee of Correspondence.[68] The Boston committee therefore sought to expand its own network by building on the commercial connections that printers had established among themselves.

The Boston committee's efforts were exceedingly successful. Just a few months after its formation, Samuel Adams wrote to his friend Arthur Lee, an American living in London, about the committee's work. He specifically addressed criticisms of the committee that loyalists had raised, arguing, "It is no Wonder that a Measure calculated to promote a Correspondence and a free Communication among the people, should awaken Apprehensions" among loyalists. The group's success in uniting the towns of Massachusetts and other colonies "must detect their Falshood in asserting that the people of this Country were satisfied with the Measures of the British parliament and the Administration of Government."[69] The cooperation of Boston's printers proved vital to this effort. Even within the local context, where oral culture was at its most powerful, resistance leaders sought out and utilized the power of the press to broadcast their message. Doing so allowed the committee to keep interest in its cause high and to demonstrate graphically the waves of support it received. At the same time, however, printers were reluctant simply to allow the committee to have its way and pushed back to force the committee to work with them to publish materials in a manner that best suited their commercial interests.

Coordinating Resistance to Tea in Print

The opportunity for printers to fully mobilize their commercial networks came in opposition to the Tea Act in 1773. These protests would prove in many ways to be the climax of printers' efforts to create an effective communications infrastructure for the circulation of political news in a local and intercolonial context. As during the Stamp Act crisis eight years earlier, printers made the protests appear unanimous and made sure that news and arguments reached across the colonies. By the summer of 1773, the printing trade had begun to take on a pronounced political cast. Unlike the Stamp Act crisis in 1765, the threat of the duty on tea had little effect on printers' main source of income (though some undoubtedly sold tea as part of a side business, and many would have consumed it). Printers therefore were much less driven to act in response to a threat to their commercial interests in 1773 than in 1765. At the same time, the political field had shifted in the intervening eight years. In 1765 the political factions that would come to be identified as Patriots and loyalists (the self-styled "friends of government") were not yet sharply defined. Most printers opposed the Stamp Act, even as their reactions ranged from tacit refusal to print to overt and loud protest. Eight years later, many printers had aligned with the Patriots or loyalists or were identified by public opinion as thus aligned.

As had happened in 1765 with the Stamp Act, the crisis over tea had its origins in imperial problems in which the colonies were only tangentially involved. In the early 1770s, the East India Company found itself beset by scandal and mismanagement, which left it with staggering debts and a backlog of tea languishing in London warehouses. These issues threatened to bankrupt not only the merchants and traders who worked for and with the company but also the politically connected men who had invested in it (including many members of Parliament). Faced with the possibility of financial disaster, Parliament came up with a plan in early 1773 to rescue the company.[70] As part of that plan, Parliament gave new life to a tax on tea sold in the colonies—a provision that has been initially passed as part of the 1767 Townshend Acts and remained in force after most of the other duties were repealed in 1770. Parliament lowered the duty on tea and stepped up enforcement, as it had done with other commodities over the previous ten years. To help the East India Company, Parliament granted a monopoly over the sale of tea in the colonies, allowing it to recoup its losses by selling off its backlog at prices cheaper than the smuggled tea that Americans had been buying.[71] At the same time, the Tea Act presented political problems for American colonists very different from those posed by the Stamp Act. They had been drinking tea for decades and had already protested in the late 1760s when it had been included among the items subject to the Townshend duties. It was the one item that continued to be taxed when the other duties were lifted in 1770. Americans took umbrage, especially at the parallels to the Stamp Act, not least the appointment of tea commissioners who would receive the East India tea for resale in the colonies. These appointments reeked of favoritism and, according to Patriots, imperial oppression.

In addition, young printers who began to rise to prominence in the early

1770s saw the trade as more inherently political than did the older printers who had trained them. When Isaiah Thomas opened his printing office in Boston in 1770 at the age of twenty-one, he joined a crowded newspaper market in Boston: his *Massachusetts Spy* was the sixth paper in the town. Like his predecessors, he dedicated his newspaper to the concept of the "free press," the idea that the newspaper was a forum for public discussion and would not outwardly espouse a partisan stance. He even added on the nameplate of the *Spy* the motto, "Open to all Parties, but influenced by none." By the time of the tea crisis, Thomas was working closely with the Sons of Liberty and the Boston Committee of Correspondence to publish anti-imperial documents. Likewise, John Dunlap, who took over his uncle William's printing office in 1768 (also at the age of twenty-one), claimed in the first issue of the *Pennsylvania Packet* that he planned to print "upon principles the most impartial and disinterested."[72] Within a few years, he too became a staunch supporter of the Sons of Liberty and anti-imperial causes. Trained during a period of great turmoil in the colonies and arriving as master printers in considerably more competitive markets, these printers had much less incentive to placate the entire populace and saw potential advantage in currying favor with a particular group.

It was in this media environment that printers heavily influenced the framing of the Tea Act as an example of imperial oppression. Through their newspapers, printers exercised enormous influence on what information circulated during the spring and summer of 1773. They distributed stories about the Tea Act, the East India Company and its monopoly in the colonial market, and the protests about the proposed three-pence duty. Their role in shaping the news was magnified because the initial information from Britain about exactly what Parliament had mandated and how it would affect the colonies was so unclear.

Because printers had developed an array of anti-imperial protest techniques and rhetorical gestures, the protests against the Tea Act mirrored and built on earlier work. In fact, printers and other protesters explicitly cited the Stamp Act crisis as a precedent for the opposition to the landing of dutied tea. When news spread of the identity of Boston's tea consignees, including the merchants Richard Clarke, Benjamin Faneuil, and Thomas Hutchinson's two sons, Thomas and Elisha, a newspaper report spread that they would find themselves hated in Boston for their perfidy. Even "men much more respected than *they* are," the report noted, would be found "as obnoxious as were the Commissioners of stampt paper in 1765."[73] Protesters were similarly interested in securing the resignations of these officials and deployed the same language and coercion techniques that they had mastered in the 1760s. In Philadelphia the *Pennsylvania Chronicle* applauded the tea consignees, who "*wisely* and *virtuously* determined to have nothing to do with so pernicious a Business, while the Teas are subject to a Parliamentary Duty, for the Purpose of raising a Revenue in America."[74] In doing so, printers tacitly demonstrated the power of protesters to convert often reluctant tea commissioners into Patriots. When commissioners, including William Kelly in New York, failed to resign immediately, the Sons of Liberty mobilized crowds to force their hand, and printers dutifully chronicled their actions. As they had during the Stamp Act crisis, printers carefully crafted their accounts of public demonstrations to portray them as orderly and lawful.[75]

Printers and their allies integrated oral, manuscript, and print media forms to bring the full force of public opinion against those who supported and abetted imperial policy. For example, many printers published letters to the tea commissioners, ship captains, river pilots, and others responsible for bringing the ships to port and landing the tea. These newspaper pieces, laid out on the page to resemble letters, operated at the nexus of public and private communication. In New York in November 1773, a letter appeared signed by "Legion" and addressed "to the STATED PILOTS of the Port of New-York, and all others whom it may concern," urging them not to aid the tea ships. "The merchants and all the inhabitants, friends to liberty," Legion continued, "are concerned in your giving the obstruction, and will support you." The writer warned of "vengeance" if any pilot attempted to navigate a tea ship to the wharves. In closing, Legion invoked the power of print as a material object, insisting that "every Pilot possess himself with a copy of this for his government" as a reminder of these warnings.[76] The publicness of the letter, both in its fact of appearing in the New York newspapers and then in its circulation to other ports, constrained the possible actions for people on the ground—just as printers intended.

At the same time, printers struggled to manage incomplete and imperfect information about the crisis. Rumors circulated throughout the fall of 1773 about the ships carrying the tea from England: how many there were, when and where they would arrive, and the identities of the ships and captains in question. Part of the problem in ascertaining this information was that ship captains with long experience in the Atlantic trade had some sense of public outrage in the colonies. Many were therefore unwilling either to transport tea or to admit it if they were. In one case, a report circulated from Boston that a Captain Scott, along with several other "Captains of vessels for this port," were "offered the tea intended to be sent" to Philadelphia, but none would take it. Scott apparently thought, given the public outcry, that the East India Company might end up not sending tea at all.[77] Similarly, William Bradford reported in the *Pennsylvania Journal* that he had received letters from London estimating that the tea ship for Philadelphia would "sail about the 12th or 15th of September" from England.[78] No one could say for sure, and newspaper pieces proposed many conflicting dates, ships, and captains. James Rivington, a loyalist printer in New York, suggested that only one tea ship would be headed to each of New York, Boston, and Philadelphia, and he published letters suggesting that "no duty will be paid upon those teas in America," eliding the distinction that Patriots had already declared in their rhetoric made no difference.[79]

The protests against the Tea Act reached their zenith in Boston, which had by 1773 developed a reputation for being in the vanguard of the anti-imperial movement, and for good reason. First, radicals in Boston were very well organized and had established a mutually beneficial relationship with several of the town's printers. Second, the colony's imperially appointed officials were particularly strong willed and often refused to diffuse tumult through compromise. In addition, since 1768 Boston had been the near-constant home of several regiments of British Army regulars. The unwelcome and unfriendly army, whose presence many Patriots felt violated their constitutional rights as Englishmen, only added to their complaints against Parliament, the ministry, and King George III.[80] For Bostonians, then, the Tea Act and its implications were only the latest in a string of increasingly unbearable insults.

Patriots largely dictated the course of events during Boston's local tea crisis. As in many other ports, the Boston Committee of Correspondence and other Patriot leaders had organized in advance to refuse the landing of East India Company tea at the town wharves. Once the first of three tea ships appeared in Boston harbor on November 27, the committee called a town meeting to organize resistance. Patriots agreed that they would hold their ground and not allow the tea even to be landed. Instead, they demanded that the tea ships return to England immediately, creating a dilemma for the ships' captains and owners. Under British law, any ship that arrived in a port had to clear its cargo with local customs officials and land it within twenty days. Colonial officials led by Governor Thomas Hutchinson refused to allow the ship to depart without landing its cargo of tea and paying the relevant duties. A standoff ensued in which no party would budge: Patriots would not allow the tea to land, officials would not allow the ship to leave without registering and landing its cargo (including the tea), and the ships' owners stood to lose enormous sums of money while their ships lay idle and fully loaded. Tension built in Boston over the ensuing three weeks as the city's protest leaders organized a series of massive public meetings and as the committees of correspondence for Boston and neighboring towns held numerous strategy sessions. To ensure that nothing happened to the tea, the Boston committee posted sentries at the wharf to guard the three ships, the *Dartmouth*, the *Eleanor*, and the *Beaver*.

Finally, on the afternoon of December 16—the last day the *Dartmouth*, which had arrived first, could legally process its cargo through customs—leaders called one final town meeting at the Old South Meeting House, creating a dramatic scene. The assembled crowd implored Francis Rotch, owner of the *Dartmouth*, to intercede with the governor and customs officials to allow the ships to leave without unloading the tea, but his attempt was unsuccessful. According to legend, shouts then came from the back of the room, which may have been prearranged signals to indicate the next move. The crowd scattered, and after dark a band of men appeared in disguises, their faces covered with soot. They marched purposefully down to the wharf, boarded the three ships, carefully opened the 340 chests of tea, and dumped the tea into Boston harbor.[81]

All of these protests, legal inquiries, and political machinations were dutifully chronicled in the city's five newspapers as printers and their political allies worked to shape the narrative of the standoff for consumption elsewhere. For Patriots, the overriding messages of the protests were their unanimity, their reasoned defense of constitutional principles, and the commendably calm comportment of the crowds assembled at the increasingly overflowing town meetings. As the tea ships arrived in late November, Edes and Gill published a note with an account of the first town meeting about the tea, noting without equivocation that "*we are assured it will not be permitted to be landed or sold here, it being the Determination of almost all the people, both of Town and Country, resolutely to oppose this artful Measure of the India Company in every possible Way*."[82]

Throughout the three-week standoff, Patriot leaders circulated news through the most effective channels possible. Shortly after the town's first set of meetings from November 29 to December 1, for example, the Boston committee dispatched an express to New York and Philadelphia with the meeting's minutes. Several newspapers reprinted the minutes, and Hall and

Sellers, publishers of the *Pennsylvania Gazette*, published them as a standalone broadside.[83] Meanwhile, Bostonians could track the progress of their reports when New York and Philadelphia newspapers provided news back to New England. Thus, Isaiah Thomas's *Massachusetts Spy* reprinted on December 16 the account of the *New-York Journal*, published by Sons of Liberty member John Holt, announcing the passage of the express rider through New York with the meeting minutes. Holt emphasized that the inclusion of the account in the *Journal* had "engaged all our people the whole night," underscoring his commitment to the cause and offering an apology to his readers for any shortcomings they found elsewhere in that issue. The feedback loops that this circulation exemplified were a crucial feature of the revolutionary news ecosystem.

Because of that process, the circulation of news of the tea's destruction, accomplished on horseback within hours, took somewhat lower priority in Boston's newspapers. Because of the publication schedule of the Boston newspapers, no accounts appeared in print until four days later, on Monday, December 20—and, in any case, nearly everyone in town probably knew of the destruction of the tea without having to read about it. The last update about the protests, therefore, came from the *Massachusetts Spy*, published on December 16 mere hours before the "Mohawks" took to the wharf. The news had already circulated, so Boston's newspapers published accounts of the tea's destruction under another mandate of the newspaper, that of historical chronicle.[84] Edes and Gill, the staunchest supporters of the Patriot cause among printers, assumed that their readers did not need access to the news quickly. As they noted in their December 20 edition, "*The particular Account of the Proceedings of the People at their Meeting on Tuesday and Thursday last, are omitted this Week for want of Room*."[85] They avoided printing the account even though they published a two-page advertisement-laden supplement to that week's newspaper. Other printers published brief accounts of the event, which circulated through ordinary channels via the post office during the rest of December and into January.[86]

Meanwhile, Boston's Patriot leaders were distributing the news through the channels constructed for precisely this eventuality, utilizing not only the stories printed in the newspapers but also manuscript letters and verbal reports. Although the "Mohawks" in Boston had hidden their identities as they worked on the tea ships, Patriot leaders—in particular, the Boston Committee of Correspondence—aimed to publicize their actions as broadly and quickly as possible. Within hours of the tea's destruction, Patriots in Boston had dispatched Paul Revere to ride south to New York and Philadelphia with the news.[87] An express rider took the dispatches from Revere in New York and carried them to Philadelphia, where they shaped how that city's Patriot leaders resisted the arrival of tea.

That port was still awaiting its tea ship when the news from Boston arrived, though it had similarly prepared to refuse the tea. Throughout the fall, Philadelphia held meetings like those in Boston where citizens pledged not to purchase or drink East India tea.[88] Philadelphia's committee likewise identified the tea commissioners and forced them to resign. In so doing, the committee and the printers who reported events used the force of public opinion as a source of legitimacy for the pressure tactics they used on the commissioners.[89] The Philadelphia committee also made efforts to prevent the tea ship from arriving, which would avoid the problem that Boston faced of what to do with the ships once they had landed. As in New York, Patriots focused considerable energy on the river pilots who guided ships into harbor, addressing several letters to them signed by the "Committee for Tarring and Feathering." The letters, which printers published in broadside form for distribution around the city and then in excerpts in newspapers around the colonies, threatened an eponymous punishment from the committee for those who aided the tea ship. One letter chillingly described what awaited the captain should he land the tea: "What think you Captain, of a Halter around your Neck——ten Gallons of liquid Tar decanted on your Pate——with the Feathers of a dozen wild Geese laid over that to enliven your Appearance?"[90] Working with printers, radical leaders circulated their threats far more broadly than would otherwise have been possible by word of mouth.

Because Philadelphia was still awaiting its tea ship, the news arrived from Boston at a critical moment. The express rider from New York reached Philadelphia on Christmas Eve, just a week after the event—about twice as fast as it typically took for news to travel between the two ports—with several accounts of what had happened in Boston. William Bradford, publisher of the *Pennsylvania Journal*, quickly printed the news in a broadside that he advertised as a "Christmas Box" for his readers—language that suggests that Bradford offered it for free to maximize dissemination (fig. 6).[91] The broadside emphasized the importance of the account by noting precisely its time of

Figure 6 (facing page). "Christmas-Box for the CUSTOMERS of the PENNSYLVANIA JOURNAL" (Philadelphia: printed by Thomas Bradford, 1773). Rare Books & Manuscripts Department, Boston Public Library

Chriſtmas-Box

For the CUSTOMERS of the PENNSYLVANIA JOURNAL.

FRIDAY Afternoon 5 o'Clock, Dec. 24, 1773.

PHILADELPHIA, Dec. 24.

A Two o'Clock this Afternoon arrived in this City a Gentleman, who came Expreſs from New-York, with the following intereſting Advices from BOSTON, which were ſent there by Expreſs alſo.

BOSTON, Dec. 16.

IT being underſtood that Mr. Rotch, owner of the ſhip Dartmouth, rather lingered in his preparations to return her to London, with the Eaſt-India Company's Tea on board, there was on Monday laſt, P. M. a meeting of the Committee of ſeveral of the neighbouring towns, in Boſton, and Mr. Rotch was ſent for, and enquired of whether he continued his reſolution to comply with the injunctions of the body aſſembled, at the Old-South Meeting-Houſe, on Monday and Tueſday preceeding. Mr. Rotch anſwered, that in the interim he had taken the advice of the beſt council, and found that in caſe he went on of his own motion, to ſend that ſhip to ſea in the condition, ſhe was then in, it muſt inevitably ruin him, and therefore he muſt beg them to conſider what he had ſaid at the ſaid meeting, to be the effect of compulſion and unadviſed, and in conſequence that he was not holden to abide by it, when he was now aſſured that he muſt be utterly ruined in caſe he did.

Mr. Rotch was then aſked whether he would demand a clearance for his ſhip in the Cuſtom-Houſe, and in caſe of a refuſal enter a proteſt, and then apply in like manner for a paſs, and order her to ſea. To all which he anſwered in the negative, the committees, doubtleſs, informing their reſpective conſtituents of what had paſſed, a very full meeting of the body was again aſſembled at the Old-South Meeting-Houſe on Tueſday afternoon, and Mr. Rotch being again preſent, was enquired of as before, and a motion was made and ſeconded, that Mr. Rotch be enjoined forthwith to repair to the Collector of the Cuſtoms and demand a clearance for his ſhip, and ten gentlemen were appointed to accompany him as witneſſes of the demand. Mr. Rotch then proceeded with the committee to Mr. Harriſon's lodgings, and made the demand. Mr. Harriſon obſerved, he could not give anſwer till he conſulted the Comptroller, but would at office hours, next morning, give a deciſive anſwer. On the return of Mr. Rotch and the Committee to the Body with this report, the meeting was adjourned to Thurſday morning at ten o'clock.

THURSDAY.

Having met on Thurſday morning, 10 o'clock, they ſent for Mr. Rotch, and aſked him if he had been to the Collector, and demanded a clearance, he ſaid he had; but the Collector ſaid, that he could not, conſiſtent with his duty, give him a clearance, till all the dutiable articles, were out of his ſhip; they then demanded of him whether he had proteſted againſt the Collector; he ſaid he had not: They ordered him* upon his peril, to give immediate orders to the Captain, to get his ſhip ready for ſea, that day, enter a proteſt immediately againſt the Cuſtom-Houſe, and then proceed directly to the Governor, (who was at his ſeat at Milton, ſeven miles off) and demand a paſs for his ſhip to go by the Caſtle. They then adjourned to 3 o'clock P. M. to wait on Mr Rotch's return, having met according to adjournment, there was the fulleſt meeting ever known, (it was reckoned, that there were two thouſand men from the country) they waited very patiently till about 5 o'clock, when they found Mr. Rotch did not return, they began to be very uneaſy, called for a diſſolution of the meeting, and finally obtained a vote for it: But the more moderate part of the meeting fearing what would be the conſequences, begging that they would re-conſider their vote, and wait till Mr. Rotch's return, for this reaſon, that they ought to do every thing in their power to ſend the Tea back, according to their reſolves. They obtained a vote, to remain together one hour longer; in about three quarters of an hour Mr. Rotch returned, his anſwer from the Governor was, that he could not give a paſs till the ſhip was cleared by the Cuſtom Houſe, the people immediately, as with one voice, called for a diſſolution, which having obtained, they repaired to Griffin's wharf, where the tea veſſels lay, proceeded to fix tackles, and hoiſted the tea upon deck, cut the cheſts to pieces, and threw over the ſide; (there ware two ſhips and a brig, Capt. Hall, Bruce, and Coffin, each veſſel having 114 cheſts of tea on board,) they began upon the two ſhips firſt, as they had nothing on board but the tea, then proceeded to the brig, which had hawled to the wharf, but the day before and had but a ſmall part of her cargo out. The Captain of the brig begged they would not begin with his veſſel, as the tea was covered with goods, belonging to different merchants in town, they told him the tea they wanted, and the tea they would have; but if he would go into his cabin quietly, not one article of his goods ſhould be hurt. They immediately proceeded to remove the goods, and then to diſpoſe of the tea.

It was expected that the men of war would have interfered, as all the Captains and other Officers were ordered on board their ſhips before night; and the day before, there were ſix dozen of lanterns ſent on board the Admiral's ſhip. The King-Fiſher, and ſeveral armed ſchooners were rigged and fitted for ſea, and the Gaſpee armed brig, arrived that day from Rhode-Iſland, But the people were determined. It is to be obſerved, that they were extremely careful, that not any of the tea ſhould be ſtolen, ſo kept a good look out, and detected one man filling his pockets, whom they treated very roughly, by tearing his coat off his back, and driving him up the wharf, through thouſands of people, who cuff'd and kicked him as he paſs'd.

We are poſitively informed, that the patriotic inhabitants of Lexington, at a late meeting, unanimouſly reſolved againſt the uſe of bohea tea of all ſorts, Dutch or Engliſh importation; and to manifeſt the ſincerity of their reſolution, they brought together every ounce contained in the town, and committed it to one common bonfire.

We are alſo informed, Charleſtown is in motion to follow their illuſtrious example.

Quere. Would it not materially affect the bringing this deteſtable herb into diſuſe, if every town would enjoin their Select men to deny licences to all houſes of entertainment, who were known to afford tea to their gueſts?

Our reaſon for ſuggeſting this, is the difficulty theſe people are under to avoid diſhing out this poiſon, without ſuch a proviſion in their favour.

We have this moment received intelligence that Mr. Clarke's brigantine, commanded by Captain Loring, bilged at the back of Cape-Cod. The Captain has not landed his Tea there, of which he has 58 cheſts on board, belonging to the Eaſt-India Company.

NEW-YORK, Dec. 22.

Laſt night an expreſs arrived here from Boſton, who left it on Friday laſt, and brings ſundry letters among which is the following, viz.

Boſton, 17th December, 1773.

GENTLEMEN,

YESTERDAY we had a greater Meeting of the Body than ever. The country coming in from twenty miles round, and every ſtep was taken that was practicable for returning the Teas. The moment it was known out of doors, that Mr. Rotch, could not obtain a paſs for his ſhip, by the caſtle, a number of people huzzaed in the ſtreet, and in a very little time, every ounce of the Teas on board of Capts. Hall, Bruce, and Coffin, was immerſed in the Bay, without the leaſt injury to private property.

The ſpirit of the people on this occaſion ſurpriſed all parties, who viewed the ſcene.

We conceived it our duty to afford you the moſt early advice of this intereſting event, by expreſs, which, departing immediately, obliges us to conclude.

By Order of the Committee.

P. S. The other veſſel, viz. Captain Loring, belonging to Meſſrs. Clark, with fifty-eight cheſts, was, by the Act of God, caſt on ſhore, on the back of Cape Cod.

---

* *By the act, any dutiable goods on board a veſſel after lying 20 days in a harbour becomes liable to the payment of the duties. The people waited till the laſt day, and in a few hours the ſhip, (to ſecure the duties then payable) was to have been delivered to the cuſtody of the man of war.*

publication to the hour—"FRIDAY Afternoon 5 o'Clock"—and by indicating that word had arrived by express, that is, a rider specially commissioned for the purpose, at two that afternoon, just three hours earlier. Printers usually dried their publications overnight, so the broadside's first readers probably had to take care not to get wet ink on their hands.

Following standard practice in editing eighteenth-century news, Bradford compiled accounts from both printed and manuscript sources to narrate the tea's destruction. The "Christmas Box" largely reprinted text from the *Massachusetts Spy* of December 16, which appeared just before the tea was destroyed.[92] It recounted the town meetings that had occurred to encourage the ships' owners to petition the governor and customs collector to allow the ships to leaver harbor without unloading their cargo. Bradford also reproduced several shorter pieces from the *Spy*: a resolution by the town of Lexington to avoid tea consumption and to burn all of the tea in the town, a query whether towns should delicense public houses that served tea, and a notice that a fourth ship headed for Boston had wrecked on Cape Cod. At the end of the broadside, Bradford published a letter from the Boston committee to New York, published there on December 22 (two days before it reached Philadelphia) that reprised the news that the tea had been destroyed.[93] In addition to publication in the special Philadelphia broadsides, each of these pieces, including the town meeting account and the Boston committee's letter, circulated broadly through newspaper networks from New Hampshire to Virginia during late December and early January.[94]

In the middle of the broadside is one paragraph that appears nowhere else. Most likely, the account either was handwritten (though no mention of such a text appears in the papers of the Boston Committee of Correspondence) or was the best recollection of the express rider upon his arrival. Its inclusion may thus represent the best instant account of the tea's destruction. After noting that the governor "could not give a pass till the ship was cleared by the Custom House," which officials would not do, "the people immediately, as with one voice, called for a dissolution" of the town meeting and headed for the tea ships to destroy the tea. One captain tried to stop the crowd but was rebuffed: "They told him the tea they wanted, and the tea they would have." This account, it seems, never appeared in any other print publication. Boston newspapers began publishing full accounts of the destruction on December 20, and by then Patriots and printers had had time to process the events and edit their accounts for broader consumption. Indeed, the account in the "Christmas Box" reads like a series of bullet points rather than a polished narrative. Yet Bradford obscured the text's tentative nature in the "Christmas Box" by integrating it seamlessly into the narrative, sandwiched where it fit chronologically between other pieces extracted from one issue of the *Massachusetts Spy*. In other words, no one in Philadelphia would have known the difference.

In editing the broadside, Bradford engaged in a process of "remediation," in which a creator navigates between two conflicting modes of presentation: the elision of any sense of mediation or editing on the one hand, and the transparent presentation of a medium as a framework for other media on the other.[95] Bradford's broadside exhibits precisely these characteristics. It creates a coherent, seamless narrative of the Tea Party with breathless, you-are-there accounts of the tea's destruction and the immediate aftermath. At the same time, the broadside is visibly hybrid, mixing printed accounts with manuscript letters and oral reports. Bradford in fact rested some of the authority of the individual pieces on the paratextual indications of their origins. The newspaper extracts and the letter of the Boston Committee of Correspondence were reliable precisely because they were properly labeled and traceable to other sources. In the same way, Bradford elided the origins of the likely oral report: it blended fluidly lest its accuracy come into question. Using editorial practices and assumptions about reader expectations standard for the period, Bradford thus created a remediated document that could serve as an authoritative account of what happened in Boston.

Bradford's editing of the accounts for publication in Philadelphia deliberately portrayed the affair in a positive light. By using the *Massachusetts Spy*, Bradford was culling news from the publication of a fellow Patriot in Isaiah Thomas, who could be trusted to portray events in a manner complimentary to the anti-imperial cause. Furthermore, the account emphasized the broad opposition to the tea's landing—not just from Bostonians but also from residents of surrounding towns. The letter to New York pointed out that the December 16 meeting included many from the "country coming in from twenty miles round." The broadside stressed the orderliness of the meetings and subsequent destruction of the tea. The accounts of Boston's town meetings read with the antiseptic regularity of meeting minutes: motions were made, seconded, and then approved. The meeting selected "gentlemen" to accompany Rotch, the *Dartmouth*'s owner, to the customs collector. And when the tea was destroyed, the protesters took special care not to harm any other goods on board the three ships. There were no unruly mobs to be found anywhere in the printed accounts. In his selection of materials, therefore, Bradford carefully amplified the representation of the Boston newspapers that the tea's destruction was the last resort of a people aggrieved by an intransigent, uncaring administration intent on imposing its will.

The quick publication and dissemination of the news proved extraordinarily helpful to Philadelphia's Patriots awaiting the tea ship's arrival. Armed with direct evidence of the potential consequences for government officials and those with an interest in the tea, Philadelphia's leaders immediately put the news to practical use. The day after the broadside's publication (i.e., Christmas Day), the long-awaited tea ship *Polly* was sighted in the Delaware River near Chester, about fifteen miles south of Philadelphia. A committee met the ship at Gloucester Point on December 26, crucially just beyond the reach of Philadelphia's customs officials. The committee requested that Captain Ayres come ashore to appear before a town meeting, where Patriots made clear that they would not allow the tea to be landed and would oppose any attempt to land it or to register with customs officials. As proof of how serious they were, they showed Ayres the accounts from Boston to demonstrate their resolve to follow the model of Boston's behavior. Ayres, unsurprisingly, "was soon convinced of the Truth and Propriety of the Representations, which had been made to him."[96]

Sufficiently chastened, Ayres agreed not to land the *Polly* and to return immediately to London with his cargo. The town meeting graciously approved various measures for Ayres and the *Polly*, including allowing him to remain in town overnight to procure supplies for the midwinter return trip to London. In addition, the meeting resolved to "highly approve of the Conduct and Spirit of the People of New-York, Charlestown and Boston" in resisting the East India Company's monopoly. These resolutions too would circulate in print, further reinforcing colonial unity. In other words, reports of the destruction of the tea circulated through information pathways that continually renewed and refreshed news that fostered connections among print readership.

A few days later, Philadelphia's newspapers began to publish accounts of the city's encounter with the dreaded tea. Printers again framed their accounts in language that emphasized the unity of the colonies in the face of parliamentary oppression. Each of the city's papers published the same basic narrative of events. The essay in the *Pennsylvania Gazette* of December 29, 1773, is representative.[97] It praised the "Unanimity, Spirit and Zeal" of Philadelphians, "which have heretofore animated all the Colonies from Boston to South-Carolina" in opposing the sale of dutiable tea. The story recounted Philadelphia's months-long preparation, reprinting a set of resolutions passed by a town meeting in October. The report also jabbed at the East India Company, noting that the *Polly* was already en route back to London to deliver the company's tea "to its old rotten Place" in the warehouses, where the tea had been stored for some time. This essay became the definitive account of Philadelphia's rebuff of its tea ship, appearing in newspapers throughout North America over the ensuing weeks.[98] As the story circulated, it reinforced themes that had been developed during the imperial crisis: the sanctity of property, the necessity of fighting oppression, the importance of virtue, and the centrality of the colonies' commercial interests.

The circulation and recirculation of this news created a feedback loop that printers continually fed by reprinting news that represented public opinion as in favor of resistance to imperial policies. Because printers selected much of their material from the publications of their fellow tradesmen, the collective weight of Patriot printers' decisions helped limit the options for all printers. Furthermore, their stated commitment to impartiality (albeit of an occasionally peculiar sort) meant that printers one would expect to be publishing with an increasingly partisan slant in fact published a confounding range of news. James Rivington, the most prominent loyalist printer in the American colonies, followed this pattern.[99] During the tea crisis he published pieces meant to cast doubt on anti-imperial political arguments. In October, for instance, he flagged and reprinted a passage from the Tea Act to suggest that because the tea was "exported *free from the payment of any customs and duties whatsoever*," it was not "an infringement of the liberties of America."[100] He also published several series of essays calling the protests and their arguments into question. Yet he also published alongside the other New York printers the letters warning the city's river pilots, addresses and meeting accounts of the New York Sons of Liberty, and the *Massachusetts Spy*'s account of the Boston Tea Party—that is, the same account that appeared in Philadelphia as a " Christmas Box." In such cases the practices of the trade frequently led even printers who opposed a political argument to reproduce and amplify coverage of its proponents and their activities.

Printers contributed their commercial networks and pathways to the effort to circulate political news and protest during the imperial crisis. The newspaper was a printed manifestation of these networks because it brought together into a single place a broad range of communications media in oral, manuscript, and print forms. Printers determined what news fit the medium, how it should be published, and when and where. As they did so, they delimited the boundaries of political debate. In the case of imperial news, printers were typically editing personal correspondence addressed to them or to a close associate—correspondence filled with myriad details about metropolitan politics. Printers did not edit the Farmer's letters for publication, but they did make choices about whether to reproduce them. The letters struck a responsive chord among colonists seeking a clear distillation of their grievances, and printers throughout the colonies sought to capitalize on this response by reprinting them. For years afterward, several printers continued to profit from the Pennsylvania Farmer through publication of the letters in pamphlet form and by printing debates about them in their newspapers. In doing so for numerous other publications during the imperial crisis, printers reaffirmed and spread the arguments and complaints of anti-imperial protesters. Similarly, printers' editorial decisions gave sanction and legitimacy to extralegal political organizations operating in the colonies. Reciprocally, printers gained valuable political and commercial connections by associating with groups such as the Boston Committee of Correspondence as they had been doing since the Stamp Act crisis.

The tumultuous years from 1767 to 1773 were crucial for the development of political communications networks in the North American colonies. Building on the connections made during the Stamp Act crisis, printers with anti-imperial leanings strengthened political communications pathways by cultivating contacts across the Atlantic, throughout the continent, and within their localities. At the same time, loyalist printers became increasingly marginalized and dependent on government support. Anti-imperial printers also began to filter for content about the imperial crisis more closely than they had before 1765. Editing a letter with political news for publication, reprinting political essays, and allying with extralegal committees all became political statements that had broader ramifications for printers' businesses. As crises grew in number and frequency with tensions heightened during the late 1760s, printers' acts as editors and compilers became a crucial means of shaping those events. By the early 1770s, their close connections to extralegal political groups was becoming considerably stronger, which enabled printers not only to shape the news but also to create their own movements to combat imperial policies they found oppressive.

The Collision of Business and Politics, 1774–1775

Over the several months after the destruction of Boston's tea shipment, colonial radicals continued to oppose the landing of tea ships in various Atlantic ports, while in Britain Parliament learned of the events in Boston and began to debate how to respond appropriately. The largest parliamentary faction by far insisted on an unequivocal statement of its supremacy over the colonies. It would be impossible to punish perpetrators whose identities, even if known among the Boston community, would be very difficult to establish in any legal sense. Instead, Lord North as prime minister put forward a series of bills that would punish Boston, the colony of Massachusetts, and by extension all of the recalcitrant colonies. Over the objection of a small minority that warned of dire consequences, in the winter of 1774 Parliament passed a series of laws that came to be known as the Coercive or Intolerable Acts. The most important of these was the Boston Port Act, which ordered the port of Boston—the largest trade entrepôt in New England—closed until the people of Boston had made restitution to the East India Company for the tea dumped into the harbor. North also recalled Thomas Hutchinson and placed the colony under a military governor, General Thomas Gage, with a new detachment of troops to reinforce those already present.[1]

When news of the Boston Port Act and the other measures approved by Parliament first reached Boston on May 13, 1774, the intelligence spread swiftly. The Boston Committee of Correspondence dispatched letters via post to towns in Rhode Island and eastern Connecticut and ordered Paul Revere to ride to Hartford, New York, and Philadelphia with the news.[2] Within days the committee already had positive responses: New Haven was the first to pledge its support on May 25. Further south, the town of Baltimore received the news from Boston just twelve days after it arrived in New England (several days faster than news ordinarily traveled) and dutifully forwarded the information to Williamsburg and Norfolk in Virginia.[3]

Printers not only spread the news via the pages of their newspapers but also used their publications, as they had in the past, to protest in other ways. Two Patriot printers, for example, altered the imagery in their nameplates to emphasize the drastic circumstances. Ordinarily, most colonial newspapers used a simple stock image to adorn their nameplates at the top of the first page: a ship, a learned figure, or a version of the royal coat of arms, replete with lion and unicorn.[4] These images would appear week after week for years. In the spring and summer of 1774, two newspapers revived an old visual device from the Seven Years' War: Benjamin Franklin's "Join, or Die" emblem. First published on May 9, 1754, in the *Pennsylvania Gazette*, the device—a snake cut into eight pieces representing the colonies—was part of Franklin's effort to create a unified response among the colonies to attacks by the French and Indians on the frontier.[5] John Holt, publisher of the *New-York Journal*, simply inserted a woodcut of the snake in the middle of his newspaper's title in several issues in July 1774, changing the epithet to read "UNITE OR DIE." Isaiah Thomas went a step further, hiring Paul Revere to engrave a snake device interwoven with his already elaborate nameplate for the *Massachusetts Spy*. He created a seamless visual in which the snake prepares to do battle with a dragon (fig. 7).

Coordination among printers and groups such as the Sons of Liberty and the Boston Committee of Correspondence increased in the wake of the tea crisis. Patriot leaders began to use their own connections to support printers' business efforts. In Wilmington, North Carolina, the newly formed Committee of Safety, in one of its first actions, helped Adam Boyd to start the *Cape-Fear Mercury*. Boyd had wanted to publish a newspaper for years in order to improve the flow of information in the colony.[6] In early 1775, as a sense of crisis heightened, a group of Boston Patriots, including both Paul Revere and the printer Benjamin Edes, wrote to John Lamb, the head of the New York Sons of Liberty, offering to share information so that "you may have it in your power to contradict the many infamous lies, which are propagated by the enemies of our country," and asking him to form a group to correspond in return.[7] Committees also coordinated with printers to ensure a flow of supplies. Even under ideal circumstances they frequently struggled to stock enough paper in colonial America.[8] Printers had to make do as best they could, which sometimes meant that the quality (and size) of paper used for their newspapers varied widely, even on a week-to-week basis. With the

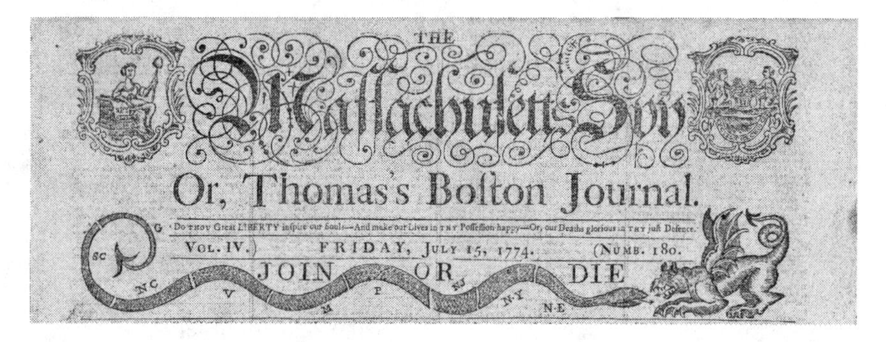

Figure 7. Massachusetts Spy masthead, July 15, 1774. American Antiquarian Society

interruption of Atlantic trade cutting off the supply of British paper, the effort became acute and urgent. Isaac Sears, a leader of the Sons of Liberty in Connecticut, attempted to help John Holt get paper from manufacturer Christopher Leffingwell through Peter Vandevoort, a New York merchant. In asking Vandevoort to intercede, Sears praised Leffingwell as "a firm Friend to the Liberties of America" and lauded Holt as the publisher of "the only Constitutional paper in this City." According to Sears, Holt always sought to publish information "that wou'd have the least tendency to promote the cause of Freedom in this Country."[9] Supplying Holt with paper, Sears implied, would be good not only for business but also for their mutual cause.

The imperial crisis and the ceaseless waves of political controversy often seemed to overwhelm all other concerns in the colonies after 1773. Yet, for many businessmen, ordinary work continued. On a day-to-day basis, their routines remained the same. People continued to make products, sell them, and buy them. Printers were no different: they continued to sell almanacs, books, and pamphlets on a range of topics. Newspapers continued to feature the arrivals and departures of commercial vessels, and their pages were still filled with advertising. Printers attempted to chase down customers in arrears—and met with as little success as at any other time in the colonial era. At various times, therefore, printers either used the imperial crisis as a vehicle for promotion or simply ignored it to continue in business with other printers and to seek customers beyond narrow ideological confines.

Printers and their customers continued to work together across the political divide. Whether or not the printer was a Loyalist, the British Army had a keen interest in speedy and reliable news, so General Frederick Haldimand, stationed in New York, requested copies of the *Pennsylvania Journal* in 1774 from William and Thomas Bradford, both members of the Sons of Liberty in Philadelphia.[10] William Strahan spent several years after his friend David Hall's death attempting to settle his account with Hall's sons, even though the beginning of the war ended most commercial contact between Britain and America.[11] Ebenezer Hazard, a bookseller in New York, in 1773 launched an attempt to print a compendium of laws on the "naval trade," for which he commissioned the Bradfords as printer.[12] He got far enough to distribute a proposal for subscriptions that included many of the major printers in New England and the mid-Atlantic region. Included were prominent Patriots such as the Bradfords, Peter Timothy of Charleston, and John Carter of Providence but also Loyalists such as James Rivington and Hugh Gaine of New York, Daniel Fowle of Portsmouth, and the Robertsons of Norwich, Connecticut.[13]

The rhetoric of the imperial crisis, which emphasized unity among the thirteen colonies, stimulated the commercial ambitions of printers, who developed protonationalist aspirations for their publications. Isaiah Thomas in Boston, for instance, sought to cultivate an intercolonial readership through the publication of literary magazines and enhance their commercial prospects.[14] By the early 1770s, fifteen printers had attempted to publish magazines, so Thomas was working well-trod ground.[15] Many of the essays and stories that appeared in the *Royal American Magazine* were reprinted from London magazines, in particular the *Gentleman's Magazine*.[16] However, Thomas presumed that Americans would submit original work: "Several gentlemen of known abilities," he suggested in his published proposal for the magazine, "have kindly promised to favour the public, through THIS channel, with essays, on various subjects for instruction and amusement."[17]

Because subscriptions were so difficult to come by for such a project, a printer contemplating a magazine had to enlist as much help as possible. Though an avowed Patriot and active in the Sons of Liberty, Thomas advertised for his magazine in numerous newspapers throughout the colonies, including papers run by Loyalists such as the *Norwich Packet* (Connecticut) and the *Boston News-Letter*.[18] He also made sure that the magazine was widely offered for sale. In the February 1774 issue, he listed nearly a dozen places where one could acquire the magazine: "Sold by D. FOWLE, in Portsmouth, New Hampshire; THOMAS & TINGES, in Newbury-Port; S. and E. HALL, in Salem; J. CARTER, Providence; S. SOUTHWICK, Newport, Rhode-Island; E. WATSON, Hartford; T. and S. GREEN, New-Haven; T. GREEN, New-London; J. HOLT, New-York; T. and W. BRADFORD, Philadelphia; A. GREEN, Maryland; R. WELLS, and C. CROUCH, in South-Carolina."[19] He thus counted among his agents printers in the colonies who were neutral or would become Loyalists, such as Daniel Fowle, Robert Wells, and Charles Crouch. These were standard practices during the colonial period, but their persistence in the face of a major intercolonial political crisis illustrates how printers distinguished between their common commercial interest as tradesmen and their partisan affiliation. Perhaps most telling, Thomas published in serial form in his magazine the *History of Massachusetts Bay*, authored by none other than the despised Loyalist, Governor Thomas Hutchinson.[20]

His ecumenical approach to the magazine business could not overcome the polarized political situation. Thomas shut down the *Royal American Magazine* in June 1774 because of the closure of the port of Boston. "*Of late*," Thomas wrote, he "*has been favoured with but few original Pieces: Fully vindicating the Propriety of the ancient Observation, that* 'Arts and Arms are not very agreable Companions.' "[21] That September, he sold the magazine to Joseph Greenleaf, a onetime justice of the peace who had worked with Thomas on several publications, including the *Spy*. Greenleaf struggled to continue the monthly schedule for the periodical. He begged his readers to submit material and tried to play into the pride of "*The sons of Harvard*." He wondered why, if they were "*disseminated throughout this continent*" and already contributing to the production of art and science, they could not sustain "an American Magazine" as "*a respectable figure in the world.*"[22] The *Royal American Magazine* ceased publication for good by the time of Lexington and Concord.

For Thomas and so many other printers, the sixteen months between the destruction of the Boston tea and the battles of Lexington and Concord thus proved to be a time of considerable tension. Business continued, and printers ran their offices as best they could, but the political disputes over imperial policies encroached further into daily life. Through this period, printers increasingly defined their roles as commercial actors through the prism of imperial politics. This chapter explores that transition. For some printers, the tension provided an opportunity to combine their political and commercial interests. William Goddard, for one, launched a crusade to overthrow the British imperial post office. Patriots had the advantage of local support in most of the colonies. Loyalists, by contrast, did not. They faced enormous pressure, including physical threats, as they navigated the difficult terrain of political minority without the same networks that Patriots enjoyed. These connections proved particularly useful when military action erupted in Massachusetts in April 1775. Using networks that Patriots had refined over a decade, printers joined with committees of correspondence and the Sons of Liberty to circulate the news from Lexington and Concord across the colonies.

An Independent "Constitutional Post"

As the consequences of the tea's destruction reverberated across the Atlantic, printers and their Patriot allies turned their attention to another pressing matter of political importance: the post office. Controlling the flow of information was critically and equally important to printers for their businesses and anti-imperial activists for their security. These needs came together early in 1774 in a movement to overthrow the British post office in favor of an American-run "Constitutional Post."[23]

The mastermind of the scheme was William Goddard, a thirty-three-year-old printer who made a habit of irritating his superiors and social betters.[24] By late 1773, Goddard was the sole operator of his printing office in Philadelphia, his onetime partnership with Joseph Galloway and Thomas Wharton having shattered in acrimony and lawsuits. In addition to his printing work, Goddard was also a veteran of the imperial post office and thus knew its limitations. His father, Dr. Giles Goddard, served as the postmaster for New London, Connecticut, in the 1740s, and William took an appointment as postmaster of Providence from 1764 to 1769. He also witnessed imperial postal operations up close as an apprentice in the 1750s to James Parker, secretary of the North American post office.

With financial support for his Philadelphia operation dwindling, Goddard prepared in 1773 to open a second printing office in Baltimore, leaving behind his sister, Mary Katherine Goddard, to operate the Philadelphia office. To ensure the success of the new office and the *Maryland Journal*, the newspaper that would accompany it, Goddard planned to use news gleaned largely from the *Pennsylvania Chronicle*. His efforts were thwarted, however, by William Bradford, the Philadelphia postmaster and printer of the rival *Pennsylvania Journal*. Goddard alleged that Philadelphia's official post rider was overcharging him and argued publicly that this "severe Indisposition" had delayed the publication of his Baltimore newspaper. He even claimed that the post rider's demands for payment were "so enormous" that he could not continue operating his press in Philadelphia.[25] Faced with the choice of paying the burdensome fees or bypassing the British postal route, he hired his own post rider between Philadelphia and Baltimore and labeled the new route the "Constitutional Post."[26] By itself this was not particularly controversial; many printers faced with similar obstacles in gathering news had done exactly that for decades. Goddard, however, had grander plans that would turn the conventional process of circulating information on its head. Already by late 1773, he wrote to John Lamb in New York that he was working on the "Sketch of a Plan . . . to give a firm Opposition to a certain unconstitutional Act of Parliamt. now operating in the Colonies."[27]

Described by one historian as a "political knight-errant always avid for new adventures," Goddard used what he perceived to be an attempt to undermine his printing business by a commercial rival to catalyze political opposition to the British postal system.[28] Without a satisfactory resolution to his distribution problem, he closed his Philadelphia shop in February 1774, ended the *Chronicle*'s run, and left his Baltimore operation in the hands of his sister.[29] From Philadelphia he then headed north, writing in the last issue of the *Chronicle* that he was "engaged in" "a Matter" "of a very interesting Nature to the common Liberties of all *America*, as well as to myself, as the Printer of a Public Paper."[30] Stopping first in New York, then in towns in Connecticut and Rhode Island, he finally arrived in Boston on March 14, where he presented a plan for a new postal system to the Boston committee of correspondence. Thus, he launched a campaign among Patriot printers and their political allies in other east coast cities to expand his grandly named "Constitutional Post" to cover the Atlantic seaboard at least as far south as Williamsburg and to serve as an extralegal alternative to the British post office.[31]

Goddard was now tapping into the broader effort to create more unified intercolonial resistance to British imperial policies. The Boston Committee of Correspondence was by this time communicating with other colonies in order to coordinate resistance to the Tea Act and was specifically seeking other issues that would coalesce colonial resistance.[32] In mid-April 1774, a subcommittee whose members included politicians such as Samuel Adams and local merchants such as Nathaniel Appleton and Joseph Greenleaf hammered out the details of a comprehensive post office plan, with support and suggestions that Goddard had gathered from committees in New York, Newport, and Providence on his journey north.[33]

After two weeks' work, the committee produced what was now called "*The* PLAN *for establishing a New* American POST-OFFICE."[34] The proposal drastically altered the business model that the post office had followed for decades. Rather than a state monopoly that existed to generate revenue for the government's use, Goddard envisioned a bottom-up institution run by and for its subscribers. His plan proposed to raise funds "for the necessary Defence of Post-Officers and Riders employed in the same" by subscription and to put the post's subscribers in charge of its management.[35] A committee of subscribers in each colony would choose postmasters, "regulate the Postage of Letters and Packets," including "the Terms on which News-Papers are to be carried," and oversee the operations of the post offices and post riders. To ensure transparency and security, the plan carefully noted that post office "Regulations shall be printed and set up in each respective Office" and that mail would be kept "under Lock and Key, and liable to the Inspection of no Person but the respective Post-Masters to whom directed." The committees would collectively select a postmaster general, who would be responsible for the finances of the overall system. In contrast to the British postal system, the plan called for a self-sufficient post office controlled by those with a direct financial stake in it—primarily printers and merchants. It also included significant protections for political correspondence and for newspapers. Furthermore, by establishing the post office as an extralegal institution, Patriots closely aligned it with political resistance groups already familiar to them, such as the Stamp Act Congress, the Sons of Liberty, nonimportation associations, and the committees of correspondence.[36]

The post office that Goddard envisioned was in all but its most basic operations modeled on the existing imperial system. The plan proposed neither new routes nor any new way to provide postal service to correspondents. As Thomas Young noted to John Lamb in May 1774, "We would not be under the least difficulty in this Colony" in making the transition from imperial to "constitutional" post, "as there would be no change in the persons employed."[37] At the same time, it marked a radical departure from the British post office. For example, it extended only within the bounds of the older colonies in North America. Neither Goddard nor the Boston committee ever contemplated a post office that would circulate information north to Canada, south to Florida, or out into the Atlantic, either to the British West Indies or to Europe. It also placed considerable power in the hands of its users, which would prevent the abuses of a heavy-handed centralized administration that Goddard felt he suffered. In particular, Goddard's plan did not attempt a grand resolution of the question of how to deal with newspapers. Instead, it charged each colony's committee with deciding how to approach the cost of newspaper distribution. Even this was a step beyond the imperial post, whose organizing statutes made no mention whatsoever of newspaper carriage.

To gather financial and public support, the Boston committee used strategies Patriots had developed over the previous decade of protests. It first contacted prominent commercial centers, towns that were already in touch with the Boston committee about the growing political turmoil in the colonies. The plan to create a new post office fit comfortably with the Boston committee's efforts to mobilize commercial towns on behalf of intercolonial resistance to imperial policies. Furthermore, the Boston committee saw the post office as one facet of its "over-all effort to see a reliable union established among the colonies."[38] The committee used various tactics to seek broad support: letters of endorsement from prominent men; travel and personal visits (mostly by Goddard himself); the use of subscription plans as an organizing device; and, when radicals sought to reach the widest possible audience, printing their arguments in newspapers.

Meanwhile, the Boston committee began the work of financing the new post office. The committee met to discuss the project twice more within a week of Goddard's arrival and appointed a subcommittee to meet with several prominent Boston merchants to begin the process of collecting subscriptions and to gather their opinions about the plan.[39] Printers and political leaders used subscription lists here to underscore support for the new post office, as they had done for nonimportation agreements in the 1760s against the Townshend Acts and in 1773 against the Tea Act. These lists were not just a crucial financing mechanism for printers and were more than registers of supporters and sympathizers. Because economic resistance depended on the cooperation of merchants, political leaders used the subscriptions lists as tools of public opinion formation, rewarding those who signed them and punishing those who refrained.[40] The subscription lists allowed printers to trumpet the patronage they received from affluent and prominent men in the town. For example, when Goddard arrived in Boston, the *Massachusetts Spy* published a series of pieces on the post office. In recounting Goddard's trials with the Philadelphia postmaster, one piece emphasized that "nearly the whole town of Baltimore, [and] the first Merchants and Gentlemen in Philadelphia," had helped him establish a post rider between those towns.[41] After Goddard's trip to New Hampshire, Samuel Cutts wrote on behalf of the Portsmouth Committee of Correspondence that "the Merchts. & Traders in this Town . . . in general esteem the Undertaking much, & are now subscribing for the Purpose of carrying it into Execution here."[42] Promoters of the new post office rarely identified individuals by name, but they referred explicitly to merchants—that is, the men whose financial backing would be absolutely necessary to the plan's success.

Patriots' use of subscription lists took advantage of merchants' ability to fund the new post office and their general concern with the flow of information. Merchants were at the center of their towns' information systems because of their need to communicate across a broad geographic range, and in many ways merchants thus exerted great control over what was considered newsworthy.[43] Merchants also had a great interest in ensuring the security and timely delivery of their correspondence. For example, when Canada came into British hands in 1760, merchants in Montreal and Quebec immediately sought the introduction of post offices in these towns to provide a secure conveyance for their business correspondence.[44] One newspaper article drafted in Quebec and printed in Philadelphia in October 1773 described the post road as "a Thing so long in Agitation, so much desired, and that must be of such mutual Advantage to both Provinces."[45] Although merchants had several options for getting their correspondence delivered, including the use of their own systems for transporting goods, they had a strong interest in a properly functioning post office. They also wanted one that protected the security of their business communications.

The most effective way for radicals to circulate their arguments on behalf of the new post office was to publish them in newspapers with the help of printers. Goddard and the Boston committee maximized their newspaper connections to generate broad publicity for the pro-American post office. Committee allies Isaiah Thomas and Edes and Gill took the lead in publishing materials in Boston. Goddard tapped into a group of expatriates from the Franklin network, most notably John Holt, the printer of the *New-York Journal.* Holt was the son-in-law of William Hunter, who had served as deputy postmaster general with Franklin until his death in 1761. Also apparently through Holt, Goddard's post office scheme reached the Williamsburg firm of Alexander Purdie and John Dixon. Purdie was Williamsburg's imperial postmaster but readily signed on to the new post office. In New England, Goddard rekindled old contacts among the Green family of printers, including Thomas Green in New Haven—also a former Holt partner—and Timothy Green in New London, as well as Daniel Fowle in New Hampshire, who was distantly related professionally to the Greens.

Newspapers deployed the same tactics as had the unpublished letters of endorsement. In fact, on occasion newspapers simply reproduced letters that Goddard carried with him or that had appeared in the newspapers of other towns when he visited them. Print coverage of Goddard's plan thus tended to follow him as he traveled, creating waves of attention both in his wake and ahead of his visits. To gain as much support as possible, the letters amplified the personal endorsement of leading men. For example, a letter dated February 28, 1774, from a "Gentleman at New York" to a friend in Boston outlining the main arguments for the new post office was reprinted in several newspapers, including the *Connecticut Gazette*, the *Connecticut Journal*, the *New-Hampshire Gazette*, and the *Virginia Gazette*. The author of that letter urged his friend to "use all your influence in the town of Boston" to gather support for the plan. A second letter printed alongside it in some newspapers noted that the post in Baltimore and Philadelphia was "supported by the most eminent merchants & other gentlemen in those places."[46]

Newspapers underscored support for the plan from around the colonies in order to project unified, positive public opinion. One article argued that support for the plan was "so universally accknowledged by the Inhabitants of this Town and Neighbourhood" that it was near certain that all the colonies would adopt the plan.[47] Another author published in Massachusetts, Connecticut, and New Hampshire newspapers noted that subscriptions had already been started in Baltimore, Philadelphia, and New York and argued that the "Southern colonies"—by which he meant Pennsylvania, Maryland, and Virginia—would quickly set up post offices once they knew that the New England colonies were interested.[48] All of these techniques—mobilizing support among each town's elite, soliciting letters of endorsement, personal visits, the use of subscriptions, and the use of newspapers—aided greatly in giving arguments on behalf of the new post office a chance to succeed.

With so many backers apparently primed for action, Patriots and printers systematically used newspapers to attempt to convince the public that creating a new post office was the right thing to do. To buttress the new post's credibility as an anti-imperial institution, Goddard and his allies made a series of arguments about the imperial post office as an oppressive arm of the British ministry. While some of these arguments were particular to the post office and to the intersection of the postal service and the printing trade, others flowed directly from the ideological opposition to British imperial policies that had developed since 1763. As a government institution with relatively broad reach across the colonies, revolutionary leaders found the British postal service a ready and potent symbol of imperial oppression and maladministration.

Proponents of the new post used three main lines of attack against the British institution. First, they argued that the existing postal system had to be replaced because it represented unconstitutional taxation, illegitimately generating revenue for the British ministry. The Boston committee put it bluntly and simply: the Post Office Act was "to all intents & purposes a Revenue Act" and the resulting institution was designed primarily to raise money for the empire.[49] Second, anti-imperial radicals contended that British officials used the post to censor their communication, arguing that the its "Officers have it in their power to intercept our communications, to extort whatever they please, and to apply them to divide us, and then to enslave us."[50] In some cases, Whig leaders had firsthand evidence that their letters were not secure. Franklin knew from his service with the imperial post that mails were routinely opened, and in 1774 he warned correspondents (including his sister) that their letters might be intercepted.[51] Finally, they suggested that a new post office, free of these constraints, would better facilitate intercolonial union. Successful opposition, they wrote, "depends, upon a free communication of the Circumstances and Sentiments of each to the others, and their mutual Councils."[52] The new post office was, in short, essential to political unity and resistance.

The "Constitutional Post" was not universally popular. Burgeoning groups of Loyalists in particular targeted Goddard's proposal as ineffective at best and treasonous at worst. In May news spread that a servant of the post rider that Goddard had hired to carry mail between Baltimore and Philadelphia named either Stimson or Stinson had stolen money from the new post office. Printers favoring the new post office attempted to dismiss the theft as an isolated event, arguing that it "by no means discourages the friends of the new institution." They also published alongside the discouraging news of the theft a piece reporting that the British postmasters general had announced the loss of almost five hundred letters from ships.[53] Over the summer of 1774, reports circulated that a group of merchants opposed the new post office plan, likely at the behest of Galloway and Wharton, who were still influential among Philadelphia's merchants and well on their way to becoming Loyalist leaders. Even so, Goddard attempted publicly to maintain his optimism, penning an essay in the *Maryland Journal* that assured Baltimore merchants that the plan was "liberally encouraged" and "nobly patronized" by merchants in New England.[54]

Whatever its success on the ground, the activity caught the attention and concern of British post office officials. John Foxcroft, the new deputy postmaster general for North America, wrote in May to London about the new post office scheme. Foxcroft at the time was contemplating an attempt to prosecute Goddard for defrauding the king of his revenue but doubted whether the charges would stick because of technical jurisdictional issues. He assured Anthony Todd, the secretary of the General Post Office, that Goddard was nothing more than a deeply indebted gadfly and that his supporters were "a Set of licentious people of desperate fortunes whose sole consequence, nay even Dependance, is on their fishing in troubled water."[55] Foxcroft's confidence obscured the impending obsolescence of the British post office in North America.

The new post office as envisioned by Goddard, the Boston committee, and other Patriots, demonstrates the importance to radicals of securing favorable mechanisms of intercolonial political communication. Because printers and radicals could not rely on the imperial post, they readily adapted the model of extralegal institutions established by the Sons of Liberty, committees of correspondence, and other resistance organizations. Although the British post officially continued to operate in the colonies until the end of 1775, Patriots had produced a systematic alternative, using fundraising and organizational techniques familiar to them from prior business and political experience. Despite their comfort in operating extralegally, however, radicals and printers never intended the post office to remain outside government control for long. They quickly began to agitate for the post office to become a governmental institution, while preserving the idea of the post as a "channel of publick intelligence" and a locus of patronage for printers.

Meanwhile, the British imperial post office was but a shell of its former self by 1775. It continued to operate, or at least attempted to, as a new infrastructure grew up around it. John Foxcroft and Hugh Finlay, who replaced Franklin as deputy postmaster general, continued through 1774 to hold board meetings at which they commissioned new postmasters and attended to the system's finances. Foxcroft even appointed a new post rider to serve between Philadelphia and Baltimore—the very route that had sparked the controversy in the first place.[56] Packet boats continued to operate between Falmouth and New York through the fall of 1775. The Continental Congress debated in October of that year whether to force the shutdown of the imperial post, and though the idea enjoyed support from the most radical members, such as Samuel Adams, Congress adopted the position of Robert Treat Paine of Massachusetts that "the ministerial post will die a natural death; it has been under a languishment a great while; it would be cowardice to issue a decree to kill that which is dying."[57] In December 1775 the last packet sailed from New York, and the imperial post office was at an end in the rebellious colonies.[58] For the next six years, communication between Britain and its former colonies could occur only through private, informal, and often technically illicit means.

The Danger of Being a Loyalist

As the complaints of American colonists grew more strident, members of the printing trade began to fracture more clearly along political lines. During the Stamp Act crisis, nearly all printers had opposed the act—it threatened their livelihoods, after all—whether or not they supported the measures used to nullify it. But by 1774 it was becoming increasingly difficult for the "friends of government" to maintain the same fiction that Patriot printers upheld, that is, that they were "meer mechanics" operating "free and open" presses for their communities. They were becoming, in other words, Loyalists.[59] Loyal printers (or neutral ones, as Patriots often denied the distinction) relied on the same types of connections as other printers, seeking out partners in government, in commerce, and within the trade. During 1774 and 1775, these links proved crucial for Loyalist printers as Patriots tried to sway public opinion against them. As part of a self-reinforcing cycle, their increasing reliance on imperial patronage more strongly associated them as Loyalists in public.

Those printers who identified as Loyalist during the imperial crisis and at the outset of the Revolutionary War formed a significant minority of the entire trade—nearly 40 percent, in fact. Like their Patriot and neutral counterparts, these printers were located throughout the British colonies. They did, however, share some characteristics in common. First, and perhaps most obviously, Loyalist printers tended to have much stronger connections to British officials and imperial governments than other printers. Many held the post of King's Printer or another imperial appointment for the colony in which they resided and worked. Second, many were recent immigrants—about a third of the overall Loyalist printers. Among these immigrants, the split was nearly even between Patriot and Loyalist, as compared to a two-to-one overall ratio. Nearly all the Loyalist immigrants (eleven of thirteen) hailed from Scotland, and eight of the thirteen had immigrated since 1763. The networks that these printers developed tended to be more insular, which is to say that most of them connected primarily if not exclusively with other Scottish immigrants. They therefore had not developed nearly as thick ties within and among colonial printers as their colleagues.[60]

Given the strong opposition to publications supportive of the imperial administration, Loyalists often distinguished carefully between business connections and politics. James Rivington, for one, seemed to thrive without a web of Loyalist connections. He continued to conduct business across ideological lines even as he became a virulent Loyalist hated by the Patriots of New York City and beyond. As late as 1774, Rivington—by then outed as a confirmed Loyalist—was still trading with the Bradfords of Philadelphia, who were noted Patriots and participants in the Sons of Liberty, ordering from them copies of an essay by John Dickinson, Thomas Jefferson's *Summary View of the Rights of British America*, and the journals of the first Continental Congress.[61] Rivington's bookselling career was about making money rather than promoting a political ideology, so much so that he wanted to capitalize on relatively popular anti-imperial political tracts. He also tried to convince the Bradfords of the commercial possibilities inherent in exploiting the political controversy. "I would advise you," he wrote, "to let a parcell of each piece produced on the Subject, lie in a conspicuous part of your Shop that the many who frequent it may be induced to purchase the arguments of the several Writers on both sides the question, which they will readily do when they present themselves, duly arranged on your compter."[62]

Rivington also sought to grow his business beyond bookselling by starting one of the most ambitious newspapers of the period, one that would prove both popular and durable, much to Patriots' chagrin. When he started *Rivington's New-York Gazetteer* in 1773, he made expansive claims for what would appear within its pages. Seeing his newspaper as having a broad geographic range, Rivington's definition of its contents went far beyond what appeared in any other contemporary American newspaper. He promised to publish

> the most important Events, Foreign and Domestic; the Mercantile Interest in Arrivals, Departures and Prices Current, at Home and Abroad, will be very vigilantly attended to. The State of Learning shall be constantly reported; The best Modern Essays, and every laudable Production from Helicon, inserted; The New Inventions in Arts and Sciences, Mechanics and Manufactures, Agriculture and Natural History, together with a regular Journal of the Proceedings in Parliament, and the Speeches, which are frequently characteristic of the Orator, in and out of Administration, shall be constantly inserted; A Review of New-Books will be included, with Extracts from every deserving Performance, each crafty Attempt with cozening Title, from the Garrets of GRUBB-STREET, shall be proscribed. In short, every Particular that may contribute to the Improvement, Information and Entertainment of the Public, shall be constantly conveyed through the Channel of the NEW-YORK GAZETTEER.[63]

Such a long list of subjects would have been ambitious for a London magazine at the time, let alone a weekly four-page newspaper published in the colonies. Few if any colonial newspapers made concerted efforts as part of their central mission to publish essays on manufactures, agriculture, natural history, and the like. These certainly appeared in newspapers from time to time but rarely as part of their core publication strategy. Furthermore, devoting half of each newspaper to advertising, as was typical in the 1770s, would have left little space for much attention to any of these priorities, let alone all of them. At the same time, Rivington was attempting to do this in an environment that was enormously hostile to the kind of political news to which he was predisposed as a Loyalist. Nevertheless, he managed to run a successful newspaper for several years, published more advertising than other newspapers, and printed it on sheets of paper larger and of higher quality than nearly all other contemporary newspapers.

Rivington and other Loyalists made the standard protestations of civility and decorum in their publications. In his inaugural *Gazetteer* on April 22, 1773, Rivington wrote in language common to most colonial printers: "No generous Mind can be delighted with the Assassination of a Neighbour's Character, and the Cause of Truth and Justice can never stand in need of illiberal and unmanly Invectives."[64] He would, in other words, seek to inform, entertain, and amuse, while avoiding scurrility. But by the mid-1770s, Patriots demanded more public circumspection from Loyalists than they did of others. The Sons of Liberty and other groups in New York City constantly pressured him, in light of his avowed loyalty to the Crown, to reassert his neutrality. Twice in the winter of 1774, Rivington issued a special statement. In January, he responded to claims from persons "*unfriendly to the Printer's interest*" that "*his press is not open to writers of different sentiments.*" He argued that he was "*ready to print the lucubrations of any author, who confines himself within the line of decorum, and is willing to be at the expence of publishing them.*" This apparently proved insufficient, so he issued a similar declaration in February.[65] Notwithstanding the protests of Patriots, Rivington continued to successfully publish the *Gazetteer* until well after the start of the war.

Publishing in a city more divided in its political loyalties than Boston, Rivington did make forays into shaping public opinion in favor of the imperial administration.[66] He was particularly active in sowing rumors about the Tea Act and frequently attempted to puncture the dominant Patriot narrative of an oppressive Parliament. For example, on June 10, 1773, he published a notice that left the impression that the Tea Act was sure to be repealed:

"Lord North lately declared, on being asked his opinion (not as a financier) that he expected the House of Commons would this Sessions rescind it." He also urged his fellow colonists to "assist in translating [the tea] into specie"—that is, to buy tea lest the East India Company be ruined. Rivington also flipped the rhetoric of protestors on its head, explaining that the anticipated measures meant that "it is now become the interest of government, to surrender to us our LIBERTIES INVIOLATE."[67] As late as November, as the tea ships approached, Rivington published a purported letter from Boston that claimed again that the East India Company would not even attempt to send tea to the colonies. His source told him "that when Administration becomes acquainted with the uneasiness it may occasion on this side of the water, a repeal of the duty, in the course of the approaching session, may be expected; and then all our hearts will be at rest."[68]

Patriots eventually targeted Rivington and intended to destroy his business, by force if necessary. In a letter addressed to Stephen Ward and Stephen Hopkins of Newport in December 1774, an anonymous group of Patriots styling themselves the "Freinds [*sic*] of America" excoriated Rivington and sought a total boycott of his business.[69] The group described Rivington as a "Pensiond Servile Wench" who was "Insulting, Reviling And Counteracting this whole Continent." They urged Ward and Hopkins to obtain a general agreement in Rhode Island not to purchase his *New-York Gazetteer* or deal with anyone advertising in it. He had to face economic sanction, they argued, lest he continue to distribute information harmful to Congress and the unity of the colonies. If Rivington continued to publish, the "Enemies of America" would distribute his invectives and false rumors more broadly. They noted that he was now acting "with the Greatest Safety" and "with but very little Prejudice to his Interest." Permitting him to operate so freely, they concluded, would delay peace, encourage the British government to deal more harshly with the colonies, and ruin the sense of "Common Cause" that the colonists felt.[70] Ironically, their argument boiled down to this: Rivington should be boycotted because he was too popular.

Like Rivington, Robert Wells of Charleston also began his American career as a successful bookseller before becoming a printer. A native of Scotland, Wells migrated to South Carolina with his young family in 1753 where he first operated a bookstore and worked as an auctioneer. In 1758 he commenced what was then the second newspaper in Charleston, the *South-Carolina and American General Gazette*, which continued under his name until 1775 (with a brief pause during the Stamp Act crisis). Like many of his Scottish brethren in the trade, Wells had stronger connections with his fellow Scots than with other members of the printing trade. For example, one biographer notes that Wells was friendly with John Stuart, the Charleston-based superintendent of Indian affairs for the southern colonies. Stuart apparently fed news to Wells on a regular basis that related to Native Americans and the frontier.[71] That particular news advantage fed Wells's rivalry with the Patriot printer in Charleston, Peter Timothy.

Though he espoused Loyalist views that led him to depart Charleston at the beginning of the Revolutionary War, Wells pursued a strategy similar to that of Rivington in New York as the imperial crisis deepened. In the *Georgia and South-Carolina Almanack*, which he published for John Tobler, for instance, Wells inserted in the 1775 edition (published late in 1774) excerpts from the votes and proceedings of the First Continental Congress, including the text of the Association, the agreement to suspend all trade with Britain and to enforce it through committee action.[72] And the early 1770s were some of the most productive years for his business. One bibliographer has noted that Wells's printing office sold more than thirty different titles in 1774, including a pamphlet version of Congress's votes and proceedings, *Observations on the Act of Parliament, commonly called the Boston Port Bill*, by Boston Patriot leader Josiah Quincy, and a selection of fiction and scientific literature.[73]

On his departure in 1775, Wells left his printing office in the care of his son John, who claimed to be a Patriot, at least until the British arrived to take Charleston in 1780. John Wells evacuated with the British in 1782 for the Bahamas, where he became the publisher of the *Royal Bahama Gazette*. Robert Wells's daughter Louisa also became active in the printing trade via her marriage to one of his apprentices, Alexander Aikman, who left South Carolina for Jamaica after the Revolution. He published the *Royal Gazette* in Kingston and served as King's Printer to the colony for several decades.[74] Unlike Rivington and some of the prominent Loyalist printers, Wells's reputation remained intact after the war, at least in the eyes of Isaiah Thomas. Having worked for Wells in Charleston for two years in the late 1760s, Thomas clearly retained respect and fondness for him. In his brief biography in the *History of Printing of America*, Thomas described Wells warmly despite his eventual Loyalism. "He was a staunch royalist," Thomas wrote, "but a good editor, active in business, and just and punctual in his dealings."[75]

It took just a few short months for the ground to shift underneath the feet of printers loyal to the Crown. Before 1774, most printers, politicians, and other public figures professed their devotion to the British monarchy or to the British system of government, even as many criticized particular actors within that system. Once those who opposed imperial policies shifted their position and began to attack British rule altogether, those who continued to support Britain could no longer shelter themselves as merely the more conservative among a diverse set of printers. Their political principles made them Loyalists and therefore targets, both commercially and politically. For them, the financial cost of adopting a broadly unpopular if principled political stance could be postponed but ultimately not avoided.

Spreading the News of War

During the fall of 1774 and the winter of 1775 tensions escalated as anti-imperial leaders began to project a more unified front against British policies. The Continental Congress met in Philadelphia in September 1774 and agreed to an economic boycott of all British goods through a series of nonimportation, nonconsumption, and nonexportation agreements.[76] Local communities around the colonies formed committees of correspondence, safety, and inspection to enforce the so-called Association and to circulate intelligence. In New England, local militias increased their training and preparation in case of a British attack. Boston in particular was the site of strain and potential conflict. The British blockade of the port continued, so goods had to be carted in overland from Salem, Plymouth, and other ports. Many residents, Patriots in particular, began to depart the city for fear it was not safe. The status quo held through the winter, but just barely.

The tensions in Boston exploded on the night of April 18 and the morning of April 19, 1775. General Thomas Gage, commander of the British troops in Boston and military governor of Massachusetts, ordered an expedition to march from Boston approximately twenty miles west to the town of Concord to confiscate stores of ammunition, powder, and other military necessities that pro-American militias had stored there. As all schoolchildren know, Patriots had developed an alarm system to rouse the countryside. Paul Revere took to the roads along with William Dawes to alert each town's militia along the route. They warned residents of the approaching British troops on their way to Lexington, thirteen miles away. When Revere and Dawes arrived there, they warned John Hancock and Samuel Adams, Patriot leaders staying there, that the troops might arrest them along their route. Revere, Dawes, and Samuel Prescott then pressed on to Concord, though only Dawes and Prescott made it. Revere was detained by a British sentry and held for most of the day.

At daybreak on April 19, the British force arrived in Lexington, where it was met by a group of militiamen on the town green. The commander of the force ordered the colonists to disperse. They refused. A commotion ensued, and shots were fired, leaving nine Lexington men dead. The British force continued its march west unopposed until it reached Concord. There another group of militia awaited the British troops, and the two sides fought a pitched battle at North Bridge just outside the town. Stymied in their mission, British commanders ordered a retreat to Boston. But the locals' alarms had worked well: by midafternoon, militia lined the road back to Boston and launched a series of ambushes that harassed the British lines as they marched east. By evening, when the main British force reached Charlestown directly across the Charles River from Boston, the militiamen had inflicted dozens of casualties. As militia gathered from towns around Massachusetts and beyond, the colonists set up a makeshift line to defend themselves and to hold the British in Boston. Word of mouth brought out the provincial troops, but the press would be needed to alert the other colonies.

News about Lexington and Concord spread with exceptional rapidity due to the effectiveness of "patriot alarm systems" and the organized political protest groups that sustained them. The alarm systems spread not only authentic news but also rumors, including British troop movements and purported slave revolts. One report on the night of April 18–19, in fact, led men in Framingham, Massachusetts, to believe that they had been awoken to deal with such a revolt of enslaved people in the neighboring town of Natick. Only after a few hours did they realize they needed to turn north toward Concord.[77] One of the first written accounts of the battles was a letter signed by Joseph Palmer, a member of the Boston Committee of Safety, dated at ten o'clock on the morning of April 19 as British troops were still on the move to Concord. The letter, now known as the "Lexington Alarm," addressed "all friends of American liberty." It bore the simple language of a military bulletin, informing its readers that "this Morning before break of day, a Brigade consisting of about 1000, or 1200 Men landed at Phips farm at Cambridge and Marched to Lexington, where they found a Company of our Colony Militia in arms upon whom they fired without any Provocation, and killed 6 Men and wounded 4 Others."[78] The letter asked committees that received it to offer assistance to the messenger, including fresh horses. The transmission of the letter demonstrates the importance of communications networks and, in particular, of the interplay among oral, manuscript, and printed forms of communication in spreading news of the fighting.

Within a few days, the dispatch traveled from Watertown, just outside

TABLE 2. *Transmission of the "Lexington Alarm" Letter, April 19–24, 1775*

Date	Time	Location	Approximate Miles from Previous Stop
Wednesday, April 19	10:00 am	Watertown, MA	
	Unknown	Worcester, MA	37
Thursday, April 20	11:00 am	Brooklyn, CT	40
	4:00 pm	Norwich, CT	21
	7:00 pm	New London, CT	13
Friday, April 21	1:00 am	Lyme, CT	15
	4:00 am	Saybrook, CT	5
	7:00 am	Killingworth, CT	16
	9:00 am	East Guilford, CT	9
	10:00 am	Guilford, CT	5
	12:00 pm	Branford, CT	8
	Unknown	New Haven, CT	8
Saturday, April 22	8:00 am	Fairfield, CT	24
Sunday, April 23	4:00 pm	New York, NY	60
Monday, April 24	2:00 am	New Brunswick, NJ	36
	Unknown (am)	Princeton, NJ	15
	9:00 am	Trenton, NJ	13
	Unknown	Philadelphia, PA	33

Note: Data compiled from Joseph Palmer, "To all friends of American liberty," April 19, 1775 [Am.606], HSP. Distance calculations are intended only as rough approximations. East Guilford is now Madison, Connecticut.

Boston, to Philadelphia, nearly three hundred miles away, far more rapidly than news ordinarily traveled (table 2).[79] At each town where the express rider stopped, the local committee convened to read the letter, sometimes to copy it, and to attest that it had received and verified its contents before the rider continued on to the next town. At the same time, the rider himself surely added details from what he learned as he rode. By the time the letter reached Philadelphia, it had become a product of corporate authorship. It included the original account drafted by Joseph Palmer and a series of signatures of committees in Massachusetts, Connecticut, New York, and New Jersey. In addition, the copy that reached Philadelphia included two other accounts of the battles. First, another express rider arrived from Woodstock in northeastern Connecticut with reports of the fighting that had occurred at Concord later in the day on April 19. The second was a report from a merchant from Plainfield, Connecticut, who had "just returned from Boston by way of Providence." He had "conversed with an Express" from Lexington who reported casualty numbers for both sides and encouraged "every man" who was "fit and willing" to head toward Boston. In just five days, therefore, the committee system spread the news of Lexington and Concord throughout New England and into the mid-Atlantic ports through a prearranged system of manuscript and oral communication networks.

Printers, meanwhile, played a vital role in amplifying the circulation of the news to their reading publics. They joined together printed, manuscript, and oral accounts to spread and begin to make sense of the events in Massachusetts. At least sixteen different publishers as far south as Williamsburg printed the alarm letter in their newspapers or in specially prepared broadsides by the end of April (table 3). In some cases, it appeared in the local newspaper within hours of its arrival in the hands of the town committee. The *Norwich Packet* was the first newspaper to publish the letter, which appeared in its April 20 edition, mere hours after its arrival at four that afternoon. Each published versions attempted to maintain some of the structure of the manuscript and included the names of signers, salutations, and postscripts.[80] Most also appended a preface that explained for readers how the letter had come to their towns. These prefatory notes attempted to lend credibility to the printed report by verifying the chain of communications that led to its publication. Many of these reports were simple. The broadside published by William and Thomas Bradford in Philadelphia, for example, stated, "*An Express arrived at Five o'Clock this Evening, by which we have the following Advices*"; similar language appeared in Lancaster printer Francis Bailey's broadside put out the following day.[81] Other printers gave more details, including the printers of the *Norwich Packet*, who noted breathlessly: "Just as this Paper was ready for Press, an Express arrived here from Brookline with the following Advices."[82] These prefaces lent authority to the reports with just a few key words and phrases, and by noting that the news had arrived by "express," the printers underscored for their readers that the letter had come through official (if extralegal) channels and was of the utmost urgency.

Furthermore, the shock of the event cut across hardening ideological boundaries as printers of all political persuasions eagerly reproduced the letter and other news about Lexington and Concord. Patriot printers thought it absolutely vital to broadcast the news of fighting as broadly as possible for two reasons. First, they hoped—along with the committees transmitting the news via channels of manuscript and oral communications—that the news would spur eligible men to travel to Boston to join the colonial effort and everyone else to lend support in other ways. Second, they spread the news as evidence of atrocities committed by the British to buttress their overall rhetorical strategy to paint the British as tyrannical perpetrators of oppression. But some Loyalists also printed this report, including the Robertson brothers in Norwich, Hugh Gaine and James Rivington in New York, and Benjamin Towne in Philadelphia. Open conflict between British troops and colonial

TABLE 3. *Publication of the "Lexington Alarm" Letter, April 1775*

Date	Publication Type	Publication Title	Printer	City
April 20	Newspaper	*Norwich Packet*	Robertsons and Trumbull	Norwich
April 22	Broadside	"INTERESTING INTELLIGENCE" (second express)	Robertsons and Trumbull	Norwich
April 23	Broadside	"The following interesting Advices"	unknown	New York
April 24	Newspaper	*New-York Gazette, and Weekly Mercury*	Hugh Gaine	New York
April 24	Newspaper	*Pennsylvania Packet*	John Dunlap	Philadelphia
April 24	Broadside		W. & T. Bradford	Philadelphia
April 25	Newspaper	*Pennsylvania Evening Post*	Benjamin Towne	Philadelphia
April 25	Newspaper	*Wöchentliche Pennsylvanische Staatsbote*	John Henry Miller	Philadelphia
April 25	Broadside		Francis Bailey	Lancaster
April 26	Newspaper	*Pennsylvania Gazette*	Hall & Sellers	Philadelphia
April 26	Broadside		[Mary Katherine Goddard]	Baltimore
April 27	Newspaper	*Rivington's New-York Gazetteer*	James Rivington	New York
April 28	Newspaper	*Pennsylvania Mercury*	Enoch Story & Daniel Humphreys	Philadelphia
April 28	Newspaper	*Virginia Gazette*	John Pinckney	Williamsburg
April 29	Newspaper	*Virginia Gazette*	Dixon & Hunter	Williamsburg
April 29	Broadside		Alexander Purdie	Williamsburg

militia was undeniably newsworthy, and this letter was the first written report to arrive in each town. New York's publishers, in fact, used its authority as a written document to justify printing it. Using similar language, both Rivington and Gaine stated that the day before its first publication (April 23) "we had Reports in this City from Rhode-Island and New-London, That an Action had happened between the King's Troops and the Inhabitants of Boston, which was not credited." Gaine did not identify exactly what form the other reports took, but in noting that "about 12 o'Clock an Express arrived with the following Account," he lent special credence to the alarm letter.[83] This may have been because the letter—drafted quickly before much information was available—gave a somewhat more dispassionate accounting of events than some of the more hyperbolic narratives that followed.

In the wake of receiving reports about the battles, men descended on the outskirts of Boston to help the nascent army lay siege to the British Army in Boston. The Massachusetts provincial congress and various committees all took steps to facilitate printing and circulating information to aid this mobilization. Samuel and Ebenezer Hall of Salem, printers of the *Essex Gazette*, brought their press to Cambridge to help print news and instructions for the gathering troops "at advice of sundry members of Congress & general Desire." There they not only began official printing for the provincial government but also started a new newspaper, the *New-England Chronicle*.[84] In Worcester, a convention of the county's local committees agreed at the end of May to urge someone to construct and operate a paper mill because of its "great Publick benefit" and promised a subscription effort to support it.[85] The committees and local congresses also conscripted the services of printers. As the Boston Committee of Correspondence had done in its first months of operation, it requested that printers publish materials free of charge. John Hancock, for example, asked Isaiah Thomas to publish a notice in the *Spy* calling absent members of the provincial congress back to Cambridge.[86] Thomas was feeling less charitable a few months later, however, when the provincial congress refused to take delivery on any newspapers for which it had to pay. Thomas complained to Daniel Hopkins, a member of the congress, that he had followed the orders of Patriot leaders and had continued to print his paper after Lexington and Concord despite the financial burden (he said he lacked subscribers, having only two hundred by his count). He also estimated that he had sent 288 newspapers to Congress and the army each week but charged only one penny per paper (a discount off the usual subscription cost). Thomas calculated that the provincial congress owed him some £31, 10s., and was nearly a month in arrears.[87]

The outbreak of war greatly disrupted news gathering. Ezra Stiles reported in May of 1775 from Newport (incorrectly, it turns out) that James Rivington had fled New York with Myles Cooper, the Loyalist president of King's College. He also believed that Joseph Galloway and the De Lancey family, leader of a major faction in New York politics, had sold out to the British cause. But, he cautioned, none of this information was reliable because "the Post is so irregular, the News so intercepted & the Prints so few, & the Coasters so much obstructed that we have no authentic News."[88] A year later, Stiles noted that newspapers in New York went weeks without being printed.[89] Nicholas Cresswell, a Loyalist, saw the lack of newspapers as a sign that "the Rascals has had bad luck of late and are affraid it should be known to the publick."[90] Finding reliable news and intelligence had become a difficult proposition.

The start of the war in the spring of 1775 had two distinct and opposite consequences for printers. On the one hand, the spread of news about the fighting at Lexington and Concord traveled through the channels they had created, and these pathways were crucial for the dissemination of the reports that shaped public perceptions of the battles. However, the outbreak of hostilities also precipitated massive disruptions to communications networks and caused a series of problems for printers. Information now frequently had to travel across lines of opposing military forces, making communication both more difficult and less reliable. In addition, many printers found themselves physically displaced by the fighting: the Patriots from Boston to its suburbs for a period of months, and the Loyalists away from Boston when the British Army departed. The war, it seemed, would prove perilous for printers.

As the imperial crisis heightened, the course of business for printers was filled with decisions that the reading public absorbed through a political lens. For Patriot printers, this confluence of political and economic interests was a boon. In defying imperial authority in their publications, they sought to generate anti-imperial public opinion to solidify resistance and tried to portray the colonies as unified—more so, in fact, than they actually were—in order to lend an aura of inevitability to their cause. Theirs was the easier path. Loyalists faced a far more challenging environment. Full-throated support for the imperial administration was, in most cases, simply impossible. Carefully practiced neutrality in political matters could keep a Loyalist printer out of trouble so long as his other business interests brought success. James Rivington, therefore, was both exemplary and *sui generis*. He excelled at developing commercial prospects that transcended the imperial crisis, surviving in business even as he published scathing attacks on Patriots and their cause.

The onset of war created a series of ruptures in printers' communications networks. The printers of Boston scattered to nearby towns, and those in New York would follow them into exile the following year when the British occupied the city. For Loyalists, the war made a tenuous situation into a perilous one. No longer content to fulminate against their publications, Patriots began to threaten these printers in order to shut down their presses. Many did so or retreated behind the relative safety of British military lines. Within months the period when printers could alternately work together with and rhetorically excoriate fellow tradesmen of different political persuasions would be a distant memory.

Patriots, Loyalists, and the Perils of Wartime Printing

The eight years of the Revolutionary War were difficult for the printing trade. After more than a decade of growth and increasing entanglement among printers as their networks evolved from commercial lifelines to the pathways of political protest, the fissures of the war dispersed printers geographically and cut them off from their networks. Maintaining commercial success became increasingly complicated as demand for printed matter dropped, except for government printing, and supply shortages hampered communications pathways and printers' ability to produce and distribute anything that came off their presses. Even with printers' traditional methods of circulation diminished, printers and their networks remained central not only to keeping lines of communication open among governments, armies, and civilians but also in shaping public opinion about the central ideological issues of the war, the outcomes of battles, and the meaning of events affecting the war in North America and throughout the Atlantic world.

During the decade-long imperial crisis, printers had mobilized their business networks into vital tools for shaping public opinion and disseminating political news and debate. Working in coordination with extralegal political organizations such as the Sons of Liberty, committees of correspondence, and the Continental Congress, Patriot printers pushed to portray the British ministry, Parliament, and colonial officials as corrupt and oppressive, to depict colonists as united in opposition, and to demonstrate the peacefulness and forcefulness of anti-imperial arguments. Loyalist printers for their part worked to counter those arguments, though their more limited business connections blunted their ability to act across the colonies.

The onset of war changed and, in many ways, destroyed those networks and their ability to function. At the very moment when these pathways became most crucial to provide information about the war and mobilize public opinion, printers instead faced enormous economic pressures, a war-induced diaspora, and other obstacles. The upheavals of the Revolutionary War shattered the political contours of the prewar communications system and forced printers, armies, and governments to scramble to maintain effective communication. Because the war forced them to clarify and publicize their political inclinations, printers became the focus of both governments and vigilante groups. Governments sought the most accurate information possible—something made more difficult by the war—and the gaps in regular order allowed vigilante groups to attempt to regulate what printers produced. By the end of 1776, politics predetermined the contours of a network—Patriots and Loyalists no longer shared contacts or commercial connections as they had before the war, not least because they had separated geographically.

Once a printer chose a side, however, politics ironically fell by the wayside, and economic concerns came to the fore, for several reasons. The strain of the war made supplies of printing materials (especially paper), capital, and solid information hard to come by. Printers across the United States struggled to produce their newspapers, almanacs, and other publications during wartime, and few managed to do so at the same level of quality they had before 1775. For the printing trade, therefore, the war had a counterintuitive effect. In order to set up shop in a given town, politics mattered more than ever. Once established, however, economic concerns overshadowed everything else.

Dislocation and Scarcity: The Business of Printing in War

The Revolutionary War upended not only the American economy overall but also the printing trade, as the onset of war diminished demand for printed material, fractured communications networks that printers had spent years developing, and disrupted the supply of printing materials.[1] Many of the personnel of the trade were directly affected. At the outset of the conflict in 1775, 100 printers were active in British North America. After years of slow but steady increase in their numbers through the colonial period, that growth flatlined as men fled their towns or simply closed up shop as demand for printing services vanished. At its minimum, the trade contracted to 93 members in both 1777 and 1778, but after the conclusion of the main fighting in 1781, the trade began to grow again. Of the 134 master printers working by 1783, only 65 had been active at the start of the war eight years earlier, meaning that the postwar trade would only slightly resemble that of the imperial crisis.

The printers active during the Revolution ran the spectrum of political affiliation. Of the 106 printers with known political leanings, 61 appeared to have been Patriots, 39 Loyalists, 4 neutral, and 2 switched sides during the course of the war.[2] Most of these printers eventually self-identified with one side or the other either through their publications, their affiliations, or their actions (e.g., those who evacuated for England with the British Army). Nearly 40 fought in the Revolutionary War, most of them for the United States, and a few rose to significant positions. William Bradford, in his mid-fifties, turned from his press and the Sons of Liberty to join the Continental Army and served several years as chairman of the Pennsylvania Navy Board.[3]

As with printers, the number of publications stagnated during the war. The number of imprints spiked sharply just before the war began as printers put out a plethora of essays, pamphlets, and broadsides about the imperial crisis but then dropped off each year during the war, not to recover its heights until after the ratification of the Constitution (fig. 8). The number of newspapers in North America similarly stagnated over the course of the war, hovering between forty and fifty until 1783, when the trade began to recover and expand once again (fig. 9). Despite the interest and need for news during the war, there was little economic incentive to start a newspaper, as prices on raw materials soared due to inflation and the capital required to operate a printing office was difficult to come by.[4]

Because of the financial obstacles, few printers sought to start new ventures, and those who did often saw them flounder quickly. Francis Bailey, for example, published the *United States Magazine* in 1779 under the editorship of Hugh Henry Brackenridge, an up-and-coming lawyer who had been a student of John Witherspoon at Princeton.[5] Yet the vagaries of the war prevented the magazine from achieving success, and Brackenridge and Bailey called it quits after one year. In his parting editorial, Brackenridge wrote that "it was hoped" at the magazine's outset that "the war would be of short continuance, and the money, which had continued to depreciate, would become of proper value." Given that in December 1779, both the war and inflation of Continental currency were running rampant, he concluded that they were "under the necessity of suspending it for some time, until an established peace, and a fixed value of the money, shall render it convenient or possible to take it up again." Ever attached to the cause of Revolution, Brackenridge also used the occasion to take a potshot at critics, arguing that only those "disaffected to the cause of America" or "who inhabit the region of stupidity" could possibly be happy about the magazine's demise.[6]

Because of the conflict, many printers struggled to maintain their prewar

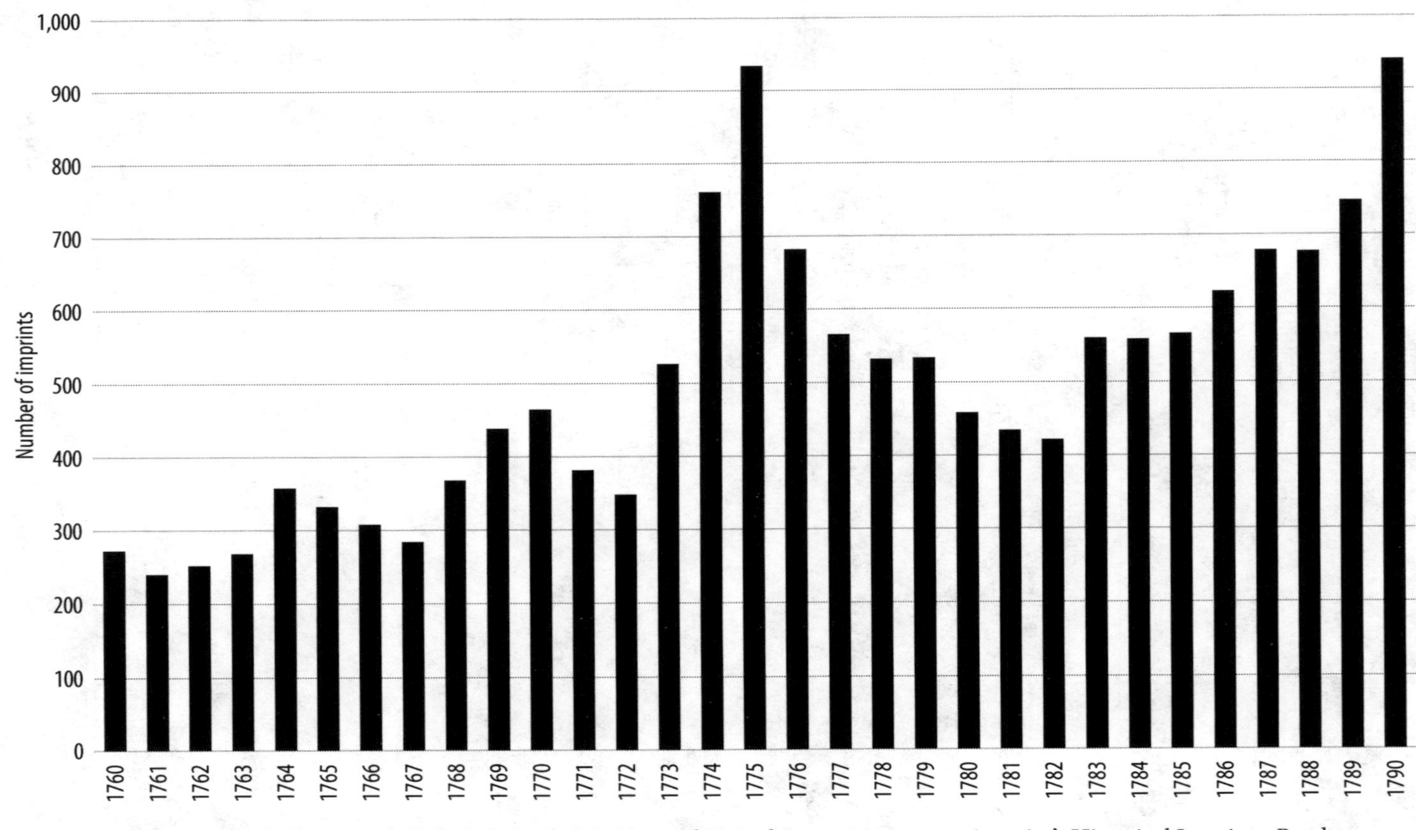

Figure 8. Imprints published in British Colonial North America and United States, 1760–1790. America's Historical Imprints, Readex

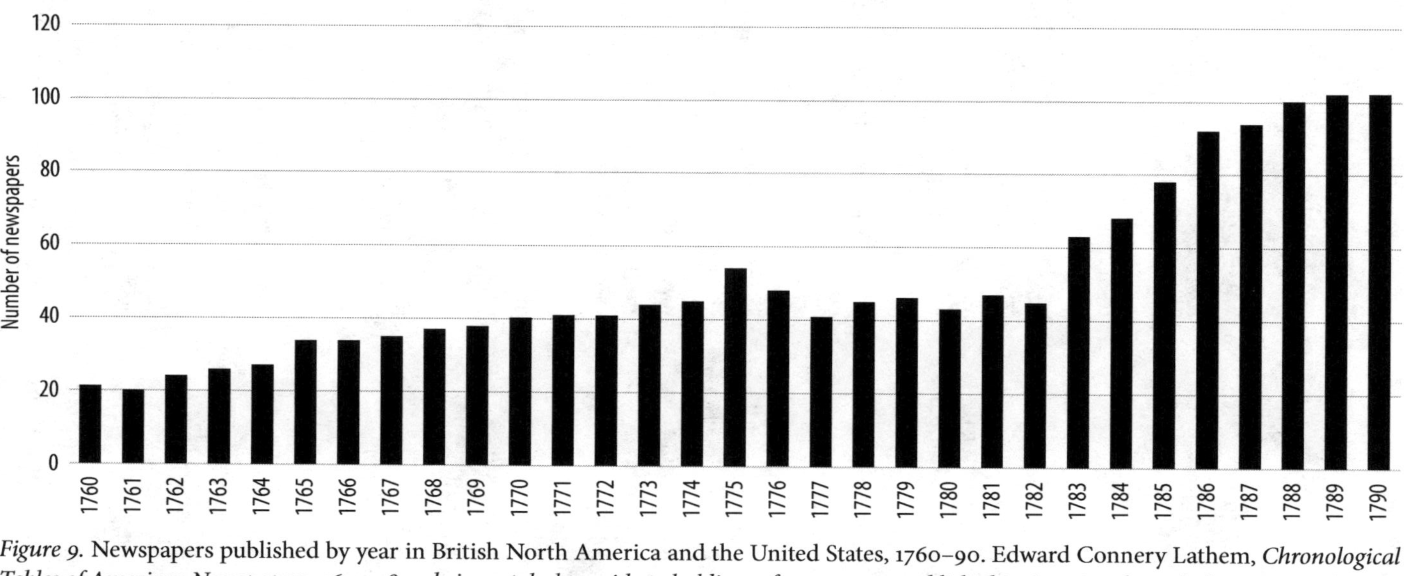

Figure 9. Newspapers published by year in British North America and the United States, 1760–90. Edward Connery Lathem, *Chronological Tables of American Newspapers, 1690–1820, being a tabular guide to holdings of newspapers published in America through the year 1820* (Barre, MA: American Antiquarian Society and Barre Publishers, 1972), 6–22; Howard S. Pactor, *Colonial British Caribbean Newspapers: A Bibliography and Directory* (New York: Greenwood, 1990). Supplemented by individual research at American Antiquarian Society.

commercial connections. Those with transatlantic connections faced particular difficulty. William Strahan, who had a decades-long friendship with printer David Hall, continued to write to his son William Hall (who was named after Strahan) after the outbreak of war. In the summer of 1775, for instance, he expressed hope that Franklin—also a longtime friend—would be able to exert his influence to bring hostilities to an end. With that letter, he enclosed "many of the new Pamphlets regarding America," lest the business relationship and its accompanying ability to profit off of controversy suffer.[7] Charleston printer Peter Timothy coordinated information with two South Carolinians in London during the war, Henry Laurens and Richard Henry Drayton.[8] When cultivating contacts, printers continued to prioritize those who could help them gain an advantage over their competitors by providing the freshest information possible. For instance, Samuel Loudon corresponded with Matthew Visscher, an Albany patriot, to get effective news about troop movements and the progress of the war, information that was "by far the best I reckon from your parts."[9]

Demand for information shifted in many ways during the war years because of the importance of military news and communication. Congress devoted considerable attention to correspondence, both the content of letters and messages especially from George Washington, other military leaders, and state officials and the means of conveying them. They passed frequent resolutions, for example, to establish express riders for the Continental Army.[10] With the British capture of New York and other locations, Congress orchestrated the relocation of specific offices: officials, for example, moved the post office some twenty-five miles north of New York City to the Hudson River town of Dobbs Ferry.[11] It also attempted to ramp up service on the coastal post road. As of late summer 1776, Congress reportedly hoped to have three posts weekly, "to ride Night & Day" with "a Rider for every 25 or 30 Miles."[12] Keeping lines of communication open was critical for delegates to ensure that news from Philadelphia could reach all parts of the new country and that information from around the United States would reach them.

Printers faced enormous challenges with sustaining private enterprises and their transatlantic connections, but demand remained in one area: government contracts. In addition to pushing printers away, each side in the conflict drew printers toward military and government headquarters to serve its needs. The Continental Congress, the rump British government based in New York City, and the fourteen state governments all required printing services for official notices and broadsides, laws, and legislative journals as well as for newspapers to promote the circulation of information and the promulgation of official viewpoints on battles, alliances, and other events during the war. For the Continental Congress and Continental Army, finding printers was typically straightforward, because Patriot printers worked in most of the places where the army was stationed. Congress sanctioned official printers in Philadelphia throughout the war and employed others when necessary, such as when Mary Katherine Goddard published the first edition of the Declaration of Independence to include the signers' names in January 1777.[13] During the British occupation of Philadelphia, Congress fled west, first to Lancaster, where it employed John Dunlap and David Claypoole, and then to York, which lacked a printer until Congress convinced the firm of Hall and Sellers to establish itself temporarily in the town.[14]

The Continental Army enlisted the services of whatever printers were nearby or used those in Philadelphia who already contracted with Congress. As militia units gathered in Cambridge in late spring 1775, Massachusetts printers took on work for the fledgling Continental Army or the provincial legislature, including Samuel and Ebenezer Hall (Salem), Isaiah Thomas (Worcester), and Benjamin Edes (Watertown). During the winter of 1778–79, the need for news was so great for the army encamped at Morristown in northern New Jersey that Congress released printer Shepard Kollock from his enlistment and provided him with a subsidy to start a newspaper, the *New-Jersey Journal*; the army may have supplied at least some of his paper.[15] The French Army under the comte de Rochambeau, stationed at Newport during 1780 and 1781, brought a shipboard printing press not only to provide for administrative printing needs but also to publish a newspaper, *La Gazette Françoise*, and an almanac for 1781.[16]

As during the colonial period, printers eagerly sought out public printing contracts to buttress the finances of their operations. Even in exile and with a war raging, Samuel Loudon and John Holt spent years vying for business with New York State.[17] States also frequently interjected themselves into aspects of the printing business beyond publishing laws, assembly journals, and official proclamations. The New Jersey legislature, at the behest of Governor William Livingston, sponsored Isaac Collins's *New-Jersey Gazette* to maintain a steady flow of information.[18] The Virginia House of Delegates went even further. When the government moved its capital to Richmond, the state bought Benjamin Franklin's printing press and type from Richard Bache for the office of Alexander Purdie, who had been given the position of state printer.[19] Politics occasionally invaded the regular commercial functioning of the trade: Yale

College, for example, sent to Hartford in 1781 for the publication of its "Commencem^t Theses, Catalogues, & Quaestiones Magistrales" because "the Press in New Haven (Tho. Green) is a Tory press & unobliging to College."[20] With opportunities for commercial gain narrowed, printers and their employers retreated to tried-and-true methods of maintaining their businesses.

The First Year of War: Trials and Triumphs

The outbreak of war in April 1775 illustrated for printers both the possibilities and the pitfalls that lay ahead of them. After a decade of publishing political tracts, essays, and news pieces during the imperial crisis, printers on both sides of the political divide began to mobilize their presses on behalf of the American and British sides. Alongside politics, printers competed to portray battles and other military events in a positive if partisan light and to circulate these accounts through the most effective channels. Most Americans had not yet begun to contemplate independence, but the chasm between colonists and Britain widened as news of the conflict spread. Nonetheless, because reconciliation remained prominent in the discussion and the prospects for a full-scale war were still unclear, printers continued to operate as much as possible within their existing networks, especially because the war was not yet affecting most of the colonies.

In Boston, by contrast, events demonstrated the difficulties that printers would face during the conflict. The outbreak of fighting at Lexington and Concord prompted a massive shake-up in the local trade. Benjamin Edes and John Gill published the *Boston Gazette* on April 17, two days before the British march from Boston. Edes then moved the printing press and other equipment to nearby Watertown, some eight miles west; the *Gazette* would not appear again until June. Once the leading anti-imperial press in New England, Edes and Gill never worked together again.[21] Isaiah Thomas moved even further, secreting his Boston printing press "out of town at Midnight by water" and moving it all the way to Worcester, more than forty miles away.[22] Adding a layer of intrigue, Thomas was able to turn the escape into a way to avoid an ordinary commercial lawsuit in the British courts; just a week earlier, one of his paper suppliers had taken out a warrant for his arrest for a debt of fifteen pounds.[23] John Boyle, another Boston printer, took his family south to Hingham for the duration of the occupation.[24] For the ensuing eleven months, little would be printed in Boston—once the publishing center of North America—save for occasional official notices from the imperial military government.

In Philadelphia, news of the battles arrived just as delegates were gathering for the Second Continental Congress, set to convene on May 10. They thus met with fresh urgency in the matter of establishing secure intercolonial political communication. Almost immediately the new Congress began discussing the viability of a Continental post office. William Goddard had attempted to convince the First Continental Congress to adopt his post office plan in September 1774, but it set aside his proposal with no recorded debate.[25] The issue therefore lay fallow through the winter and spring. After Lexington and Concord, plans for post offices proliferated from the ground up in the absence of a directive from the Continental Congress. Goddard himself immediately rekindled a New York-to-Philadelphia route, several New England colonies started provincial postal services, and numerous individuals, including several onetime postmasters and post riders, offered their services to the public.[26] Calls also began to pour in for the Second Continental Congress to take over postal service. The resolution creating a post office in Massachusetts included a clause stating that its provisions would remain in effect "until the Continental Congress, or some future Congress or House of Representatives of this Colony, shall make some further Order relative to the same."[27] With explicit support from provincial congresses and a heightened need for military communications, Congress at last seemed ready to act.

By the time it did, the landscape had changed vastly from its meeting the previous fall. Boston was occupied by British troops and under siege by local militias. A coordinated intercolonial postal system was now a military imperative. Within three weeks of convening, Congress appointed a committee to examine the postal system and then adopted its report at the end of July to establish the Continental post office.[28] Congress appointed Franklin as postmaster general, which ensured some continuity in the post office's leadership. The committee's report no longer exists, but the new congressional post was different in at least two respects from Goddard's proposal. First, the Continental post office (like its imperial predecessor) was designed on the model of a government institution, which meant that it no longer anticipated operating via subscription. Second, Congress insisted that the post be a quasi-national infrastructure that operated from Maine to Georgia; Goddard never envisioned that his post would extend further south than Williamsburg. Congress wanted a more comprehensive postal system.

Printers and others connected to the post office fully expected that the new national infrastructure would continue to rely on the patronage practices of the old imperial system. Franklin, as the new postmaster general, apparently agreed. He appointed to the new Continental post office printers and former postmasters associated with the old imperial system. For example, Newport printer Solomon Southwick held the postmastership in Rhode Island, whereas John Carter, the printer of the *Providence Gazette*, shifted his service from the British post office to the American one in Providence.[29] Alexander Purdie, who printed one edition of the *Virginia Gazette*, also shifted his duties from the imperial post to the new American post in Williamsburg.[30] Mary Katherine Goddard, who ran the Baltimore printing office of her brother William while he was traveling the colonies, was named postmistress of Baltimore. She would hold the job until 1789, when she lost her position to a new postmaster more closely connected to the new federal postmaster general.[31] Patronage positions in the post office thus continued to be vital to printers.

The printers most centrally involved in the "Constitutional Post" also sought positions but often encountered difficulty because they were perceived, perhaps ironically, as too financially interested in the post to serve the public effectively. In anticipation of action by Congress, John Holt openly campaigned for the job of New York City postmaster in an essay in his own newspaper:

> Many Gentlemen among the most hearty and able Friends to America, in this and the neighbouring Colonies, both in and out of the Continental Congress, having encouraged the Subscriber to hope, that they think him a proper person to hold the Office of Post Master in this Colony, with the Business of which he is well acquainted, and will favour his Application for the same; He humbly requests the Favour, Concurrence, and Assistance of the Honorable Convention of Deputies for this Colony, in his Appointment to the said Office, the Duties of which it will be his constant Care to discharge with Faithfulness, and to general Satisfaction, ever grateful for Favours conferred, and studious to deserve them.[32]

Holt crafted his essay as a careful appeal to the provincial congress for the local postmaster's position, referring to his own knowledge of the post office as a printer and businessman. Furthermore, like the letters of support for Goddard's plan, Holt wrote that his application had been endorsed by numerous prominent men in several colonies. Holt made his application through the public venue of a newspaper, relying on the premise that the provincial congress would hear of his interest in the job through public discussion. Even as he was applying for the official position, Holt was also involved in an informal post operating between New York and Philadelphia.[33] Amid the *rage militaire* of 1775, however, Holt's open lobbying appeared insufficiently patriotic. Thomas Bradford, William's son and partner, scolded Holt (and, by extension, Goddard) for thinking of the post office in proprietary terms. "We always tho't the post belonged to the public & not to you," he wrote to Holt the day after Congress approved the post office, for "we tho't we were serving the Public, instead of one or two private people."[34]

Also mounting a failed campaign for a position in the post office was Goddard himself. For a time he was convinced that he might be appointed postmaster general and consoled himself when Franklin snagged that position with the sense that surely he was entitled to an appointment as secretary and comptroller, the second-ranking position in the post office. That job carried a 340-dollar salary, which would have alleviated the severe financial hardship he claimed he suffered as a result of his work on the earlier plan.[35] Congress left the decision to Franklin, who instead chose his son-in-law Richard Bache. Goddard was named surveyor and charged with traveling throughout the colonies to set up post offices, just as he had done the previous year. Despite losing out on his chosen position, he set off on his task immediately.[36] He would serve the post office for only about eighteen months, however. In June 1776, he petitioned Congress for a place in the army, where he thought he could better recoup the financial losses he claimed to have suffered since 1774.

Americans also worried about postal security, not least because many postmasters had simply transferred from the defunct British postal system. To secure the mails and ensure that the post office was fully staffed, Congress exempted postal employees from military service.[37] At the same time, even the appearance of anyone unfaithful to the American cause was problematic for ensuring that letters would not be intercepted by the British. Congress therefore tried to expunge "persons disaffected to the American cause" who might have gotten positions in the post office.[38] In particular, Congress worried about the possibility of letters being stolen or transmitted to the British, potentially embarrassing the American cause or, worse, compromising a military campaign. Such concerns were well founded; both the British and Americans made a habit of attempting to intercept enemy correspondence to gain crucial intelligence.[39] Such publication could embarrass the parties involved if proved to be authentic. John Adams in the summer of 1775 had two letters intercepted by the British when the courier he had employed to carry them from Philadelphia to Massachusetts was captured. The two letters, one to his wife Abigail and the other to political leader James Warren, detailed his thoughts on independence and his disdain for fellow delegates. British officials rushed the letters to General Thomas Gage in Boston, who passed them on to Margaret Draper, a Loyalist printer, who dutifully inserted them into the *Massachusetts Gazette*.[40] They appeared in London in the fall and then back in North America in January 1776, circulating in both print and manuscript. The time lag provoked a far different response in North America than intended, because by early 1776 many Americans saw independence as a far more viable option than they had just months earlier.[41]

At the same time, Congress and the Continental Army had trouble solving one of the most intractable difficulties of eighteenth-century communications: overland travel in the South. For decades, postal officials had struggled with how to provide cost-effective service across the sparsely populated colonies marked by vast distances and difficult river crossings, whether via the imperial post office or Goddard's "Constitutional Post."[42] And in an era when oceangoing travel could be as fast as or faster than by land, North Carolina was effectively the most distant link in the Atlantic communications system, far from the mid-Atlantic ports of New York and Philadelphia and the southern entrepôt of Charleston.[43] Even toward the end of the war, transportation was slow in the Carolinas. In early 1783, North Carolina's congressional delegates urged Governor Alexander Martin to ensure that roads were under proper repair and that the state was not overcharging for ferry crossings.[44] The delegates feared that, "if the post should suffer or should be impeded by the neglect of Government—he doubtless must change his rout or be absolutely discontinued."[45] In other words, the state government had to step up to support transportation infrastructure, or North Carolina would lose access to information.

In addition to their work with the post office, printers amplified the political arguments that pushed the colonies toward independence. After the Battle of Bunker Hill in June 1775, military activity was limited for the rest of the year, though Boston remained under siege and tensions ran high in many other parts of the colonies. A *rage militaire* swept across the continent as thousands rushed to join the Continental Army or local militia units, and many headed to Boston.[46] Congress asked the king for peace in the Olive Branch petition. King George III summarily rejected its request, instead declaring the colonies in open rebellion. Even with that slap in the face, by the end of 1775 most Americans were still discussing possible terms for reconciliation.[47] That would change in January 1776.

That month, a little-known former excise collector and women's corset maker published one of the most sensational and effective pieces of political propaganda in the English language. Thomas Paine had arrived in Philadelphia from England barely a year and half earlier. He carried with him a recommendation from Benjamin Franklin and hopes of breaking into the world of publishing in Philadelphia. With the endorsement of such a leading figure, he secured a variety of jobs writing and editing for the Philadelphia political and literary scene, including a stint in 1775 editing the *Pennsylvania Magazine* for printer Robert Aitken. Paine and Aitken used the magazine to express disillusionment with the strength and breadth of the push for reconciliation with Parliament. Over time, the essays Paine wrote for the magazine grew more strident. During his editorship, for example, several pieces focused on military preparations, including an "Account of the Manufactory of Salt-Petre," a necessary ingredient in gunpowder.[48] In the July 1775 issue, he published an allegorical tale under the pseudonym "Curioso" about the "military character of ants," in which the author recounted a battle between brown and red ants. Making the allegory plain, Curioso concluded with a direct analogy between unguarded ant colonies, the security of individual homes, and the colonies of North America. One would not "tempt a thief by leaving our doors unlocked," he argued, and so "we ought not to tempt an army of them by leaving a country or a coast unguarded."[49] No longer was it possible, Paine suggested, for the colonies to count on a purported reserve of goodwill and brotherhood with England. They must be prepared to defend themselves actively.

In early January, Paine worked with another Philadelphia printer, Robert Bell, to publish *Common Sense*. In seventy-seven pages, Paine sought to shift the terms of political debate in the colonies. Unlike many of his contemporaries, Paine framed his rational argument around an emotional plea to readers he addressed as equals. The result was a grab bag of arguments that linked familial allusions, the Bible, British history, commerce, and republicanism.[50] The push for reconciliation, he argued, was folly, doomed to failure because the British would never accept the colonists as equals. Furthermore, even if reconciliation were possible, it was no longer desirable, because the British government was grounded in lies and myths, down to the monarchy itself. Most famously, in the second edition of the pamphlet Paine described William the Conqueror as a "French bastard" whose "landing with an armed banditti and establishing himself king of England against the consent of the natives, is in plain terms a very paltry rascally original."[51] On the basis of that foundation—the folly of reconciliation with a monarchy that was not even legitimate—Paine turned to his clearest argument: the commonsense move toward independence. Until that occurred, he concluded, "the Continent will feel itself like a man who continues putting off some unpleasant business from day to day, yet knows it must be done, hates to set about it, wishes it over, and is continually haunted with the thoughts of its necessity."[52] Paine, a recent immigrant unfamiliar with the divisive politics of intercolonial interactions, foresaw a future in which the colonies stood both independent and united. He therefore projected an image of a potential unified nation that captivated readers at the time and since.[53]

Paine laid out the clearest, most effective argument for political separation from Britain that colonists had yet encountered. What truly differentiated *Common Sense* from other pamphlets, however, was the effectiveness of its circulation through the colonies. Within months, twenty-five editions appeared, published in thirteen towns in seven colonies. That figure dwarfs the reprinting of any other pamphlet during the Revolutionary era. Yet the extent of the pamphlet's diffusion has remained a subject of debate among historians. Paine contributed to the problem by claiming that more than 100,000 copies of the pamphlet appeared in 1776—a number so inflated it would have been nearly impossible for printing offices in the colonies to have produced so many.[54] In reality, the extent of publication was more complicated. Of the twenty-five editions, for example, sixteen appeared in Philadelphia alone, the product of a dispute between Bell and Paine and fierce competition in the printing trade. And only a single edition was published in the South, in Charleston.[55] On the other hand, printers advertised the pamphlet for sale in other markets, such as Providence (fig. 10), or excerpted it in newspapers, including the *Connecticut Courant*, the *Norwich Packet*, and the *Virginia Gazette*.[56]

In addition to the pamphlet itself, printers circulated reflections on *Common Sense*, responses, speculation on the identity of the author, and a rousing debate about the wisdom of independence. However many copies were printed, the argument that independence was "common sense" circulated widely all over the Atlantic world and across the political spectrum. General Horatio Gates, stationed with the Continental Army in Cambridge, Massachusetts, called it "an excellent performance" and speculated (wrongly) that "our friend Franklyn [*sic*] has been principally concerned in the composition."[57] The pamphlet sold well: a bookseller in Annapolis requested that Thomas Bradford send him "three Or four dozen Pamphlets of Common sence," and printer John Carter in Providence forwarded five hundred copies to the Continental Army headquarters in Cambridge, Massachusetts,

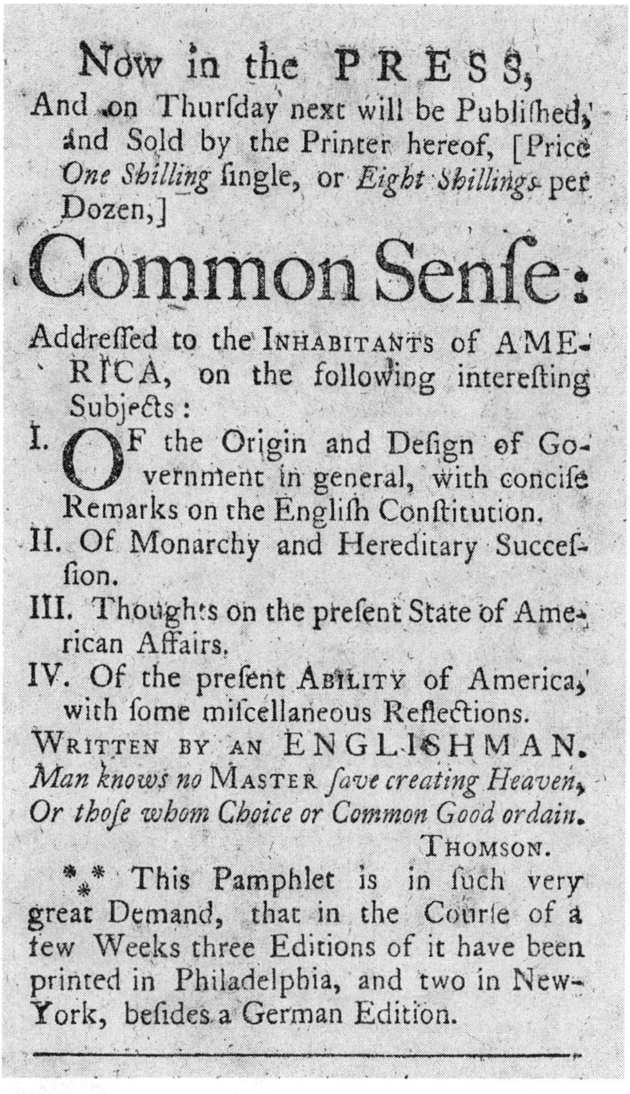

Now in the PRESS,

And on Thurſday next will be Publiſhed, and Sold by the Printer hereof, [Price *One Shilling* ſingle, or *Eight Shillings* per Dozen,]

Common Senſe:

Addreſſed to the INHABITANTS of AMERICA, on the following intereſting Subjects:

I. OF the Origin and Deſign of Government in general, with conciſe Remarks on the Engliſh Conſtitution.

II. Of Monarchy and Hereditary Succeſſion.

III. Thoughts on the preſent State of American Affairs.

IV. Of the preſent ABILITY of America, with ſome miſcellaneous Reflections.

WRITTEN BY AN ENGLISHMAN.

Man knows no MASTER *ſave creating Heaven,*
Or thoſe whom Choice or Common Good ordain.
THOMSON.

⁂ This Pamphlet is in ſuch very great Demand, that in the Courſe of a few Weeks three Editions of it have been printed in Philadelphia, and two in New-York, beſides a German Edition.

Figure 10. Providence Gazette, February 17, 1776. American Antiquarian Society

in March.[58] Within a few weeks after they arrived, a report circulated that people in Cambridge were offering a toast: "May the independent principles of COMMON SENSE be confirmed throughout the united colonies."[59] By late spring, both *Common Sense* and *Plain Truth*, the most popular response, were circulating in London from the press of John Almon, a radical London publisher who regularly served as a conduit for anti-imperial arguments from the colonies.[60]

Opponents of independence clearly believed that *Common Sense* changed the conversation on the basis of the virulence and volume of their responses. Nicholas Cresswell, a Loyalist in Virginia, described it in mid-January as "Full of false representations, Lies, Calumny, and Treason whose principles are to subvert all Kingly Government and erect an Indepen[d]ent Republic."[61] The pamphlet prompted a flurry of responses in print. Numerous writers attempted to rebut Paine's argument in pamphlets or newspaper essays of their own, most notably James Chalmers, who published a pamphlet somewhat cleverly entitled *Plain Truth* under the pseudonym Candidus, and William Smith, the provost of the College of Philadelphia, who published several essays in Philadelphia newspapers under the pseudonym Cato.[62] Robert Bell, happy to capitalize on a dispute in print, published one edition of *Plain Truth* (to which he attached an essay he had composed on the freedom of the press).[63] In their responses to *Common Sense*, Loyalist authors lavished considerable attention on Paine's role as the author, which was an open secret on first publication and which Paine all but confirmed himself within weeks of the pamphlet's first appearance.[64] Though occasionally reprinted (*Plain Truth* had four Philadelphia printings, plus two in London and one in Dublin), none of these arguments had the rhetorical force of *Common Sense*, nor did they capture the American zeitgeist in the same way.[65]

The publication of *Common Sense*, as well as other pamphlets and news stories in early 1776, made independence seem both legitimate and possible.[66] As Congress moved toward independence in the late spring of 1776, its members thought it necessary not only to formalize the states' new status with respect to Britain but also to publish a document announcing it and laying out their reasons—that is, a Declaration of Independence.[67] Though the document would later become "American scripture," to borrow Pauline Maier's phrase, in June 1776 it was simply one among many tasks that Congress undertook as it oversaw a fledgling group of independent states and a war effort against a far stronger imperial power.[68] Initially drafted by Thomas Jefferson, then edited by a committee that included John Adams and Benjamin Franklin, and finally revised by Congress sitting as a Committee of the Whole, the Declaration laid out the argument for when and why a people may rebel against a government, why and how King George III and Parliament had created those conditions, and the formal declaration by Congress that the thirteen colonies were, "and of right ought to be, free and independent states."[69]

As Congress was voting on the final text on the morning of July 4, it transmitted the Declaration to Philadelphia printer John Dunlap with an order for copies to be printed and distributed.[70] The first publicly available text of the Declaration included the signatures of John Hancock as the president of Congress and Charles Thomson as its secretary. Only these three men—Hancock, Thomson, and Dunlap—found their names on the document on July 4. The Dunlap broadsides traveled quickly through the thirteen newly independent states to each state's government, the Continental Army, the Congress's representatives in Europe, and other locations. Domestically, the Declaration appeared in at least thirty different newspapers in nine states by August 2, including a German translation in the *Pennsylvanischer Staatsbote* (July 9).[71] About six weeks after its publication, the Declaration appeared for the first time in London newspapers on August 14 and had reached half a dozen cities in Continental Europe by October.

Most Americans today think of the Declaration as a faded parchment document secured in bulletproof glass at the National Archives, viewable only on a visit to Washington, but at the time Congress clearly intended the Declaration to be a printed, public statement about its intentions with respect to Great Britain. After all, it would be of little use to declare something without an audience.[72] By printing the Declaration and ordering it read and reprinted, Congress thus distributed the news to Americans in a way that "collaps[ed] the physical distance" that separated the two groups.[73] The well-developed channels that printers had built over a decade of the imperial crisis, in other words, proved useful and helpful at precisely the moments and in exactly the ways that Americans and their leaders needed. The circulation of *Common Sense* and the Declaration in 1776 proved to be the apex of effective circulation during the war. In most of the combatant colonies, business in printing offices had continued, albeit at a higher state of alert. Printers shared news reports about fighting, the movement of the British and Continental armies, political debates in Congress, and the reaction of British officials in London. They also kept printing all manner of nonpolitical material, from spellers and music books to blank forms for merchants. The illusion that business would continue, however, would not last long.

Tribulation and Opportunity in a Drawn-out War

Between the summer of 1776 and the end of the war in 1783, printers faced increasing challenges to their businesses as military campaigns fractured American communications. The Continental Army and the British Army engaged for years in a long series of pursuits across eastern North America. The earliest phase of the war focused in New England, where British troops had been stationed since the late 1760s because of unrest in Boston. After Washington and the Continental Army forced them to evacuate in March 1776, the British launched an invasion of New York City in the summer of 1776. By September, the British controlled the city and its harbor, which they would hold until a full peace treaty was ratified in late 1783. Around the same time, the British took the lucrative seaport of Newport, Rhode Island, occupying it for more than three years. The main thrust of the campaigns then shifted to the mid-Atlantic, prompting two years of fighting in New Jersey and Pennsylvania that culminated in the British occupation of Philadelphia from September 1777 to May 1778. Beginning in 1779, the British abandoned the mid-Atlantic in pursuit of a southern strategy, hoping to capitalize on the large numbers of Loyalists they assumed would rally to fight in Georgia and the Carolinas. As such, the British invaded and occupied first Savannah in 1779 and then Charleston in 1780, ports that (like New York) they would relinquish only at the conclusion of peace negotiations.[74]

Wherever they located offices, printers faced great difficulties with supply during the war. As members of a trade that relied on the circulation of both ideas and materials, the collapse in Atlantic trade after 1775 deeply affected printers. Because of the war's uncertainty and disruptions, paper money depreciated rapidly, and bills of exchange became hard to obtain. This destabilized the economy between the extreme inflation and the difficulty of buying supplies for one's business and selling goods from that enterprise.[75] Printers had long struggled to keep up with supplies of paper, ink, and other items, as well as to obtain fresh type and a useable press. The war exacerbated these concerns, especially because of the narrow margin in which printers typically operated. First, Americans imported a considerable proportion of their paper (especially the finer varieties) from England before the war, and that supply was, for obvious reasons, off-limits. Second, the production of paper in North America collapsed during the Revolutionary War, not only for economic reasons but also because many of the workers at paper mills went off to war and the supply of rags dwindled. Even finished paper could be put to other purposes during the war, including as wrappings for shot. Furthermore, the difficulties in attracting specie meant that Congress and the state governments competed with printers for paper so that they could print currency, even as that too devalued rapidly with massive inflation.[76] Even if they could find paper, American printers had difficulty maintaining supplies of type, which were almost exclusively imported, and keeping their presses in good repair—both of which were difficult in peacetime. Those printers who had to evacuate usually left behind a significant portion of their printing materials.

The struggle for paper was real. To cope, printers tried to adapt their long-standing practices to acquire materials through their networks. In 1777, Samuel Loudon, exiled from New York City to the town of Fishkill, nearly seventy miles north on the Hudson River, lost his usual supply chain for paper from mills around Philadelphia because of the British Army's presence. On the basis of information he received from fellow exiled New York printer John Holt in nearby Poughkeepsie, Loudon dispatched a friend to New Haven with fifty dollars to try to acquire ten reams of paper from bookseller Isaac Beers. Beers was not his only option, but Loudon feared for his press should his queries fail: "I have wrote to Milton near Boston, but have got no Answer; if I can't get supplyed from boston [*sic*] or from you, I must stop—for I don't know that a supply can be had nearer."[77]

The economic challenges that convulsed American society thus manifested in the physical objects that printers produced. The ebb and flow of a printer's fate, in other words, is visible today in the archives that hold the physical copies of their newspapers. For example, the American Antiquarian Society holds a lengthy run of the *Maryland Journal*, the Baltimore newspaper started by William Goddard in 1773 and continued by his sister, Mary Katherine Goddard, during the Revolutionary War while she was simultaneously serving as the postmistress of Baltimore.[78] Through July 1776, the newspaper appeared in a weekly four-page format as it had for several years, with a masthead that included a woodcut and a motto derived from the Roman poet Horace.[79] Over the next four years, however, the newspaper as material object changed numerous times. At various points, Goddard reduced the *Journal* from four pages to two. The masthead changed, dropping the more expensive woodcut image. The size of the paper varied, sometimes considerably smaller than what was common for newspapers of the era. The quality of the paper shifted. As one flips from issue to issue, one passes from sturdy pages through to more brittle ones. Numerous issues were printed on a thick blue paper ordinarily used for making cartridges for bullets rather than printing. Goddard at various times skipped weekly issues, which is apparent because issues continued to be numbered consecutively.[80] She offered only a few hints in the pages of the newspaper as to why these changes occurred. Like other printers, she at times made requests for rags. In April 1777, for instance, she posted under the Baltimore news heading an announcement: "The Paper on which this Journal is printed was manufactured at a Mill lately erected in Elk-Ridge Landing. To supply it with Stock (which is much wanted) the Printer will give the highest Price for clean Linen Rags, old Sail Cloth and Junk."[81] The following year, Goddard underscored the need for paper by posting on the paper's masthead that the mill needed rags.[82] By early 1780, the newspaper was back on track with weekly issues printed on high-quality paper. Goddard was far from the only newspaper printer to experience these types of shortages and difficulties with paper supply.

Getting access to paper and ink was a minor inconvenience compared to the dislocation that the two armies forced on many printers. For the most outspoken partisans of each side, the arrival of an opposing army posed a direct threat to their personal liberty and exposed their persons, families, and property to violence from opponents. On top of that, the overtly partisan political atmosphere meant that even in the absence of violence or the threat thereof, commercial success often proved impossible so long as the opposing army was in town. Finally, the war forced most printers to finally take sides—something most had avoided doing overtly for much of the imperial crisis. For many printers, therefore, relocation was a defining feature of the war years as they moved away from opposing armies. In total, fifty-one printers (representing thirty-nine offices) had to evacuate at some point during the war, including twenty-one Patriots, twenty-six Loyalists, and four others (map 3).

When motivated by military action, Patriot and Loyalist evacuations followed very different patterns. Patriot printers for the most part fled into the hinterland only upon the arrival of the British Army in their town, and most returned to their original locations shortly after the British departed, as Boston's Patriot printers did after March 1776. Similarly, printers John Holt and Samuel Loudon fled New York in the fall of 1776 as the British stormed the city and headed north to towns along the Hudson River. In the fall of 1777, when the British took Philadelphia, Patriots headed west to Lancaster, York, and other interior towns. Some of them followed the Congress in order to continue government printing.[83] In nearly all of these cases, Patriot printers returned home shortly after the British departed, surveyed the damage, and resumed business.

In places where there was little military action, most printers remained in place throughout the war. For many, fighting simply never reached them. For those who were either neutral or aligned with the Patriots, there was little incentive to go anywhere unless the British Army approached. In addition,

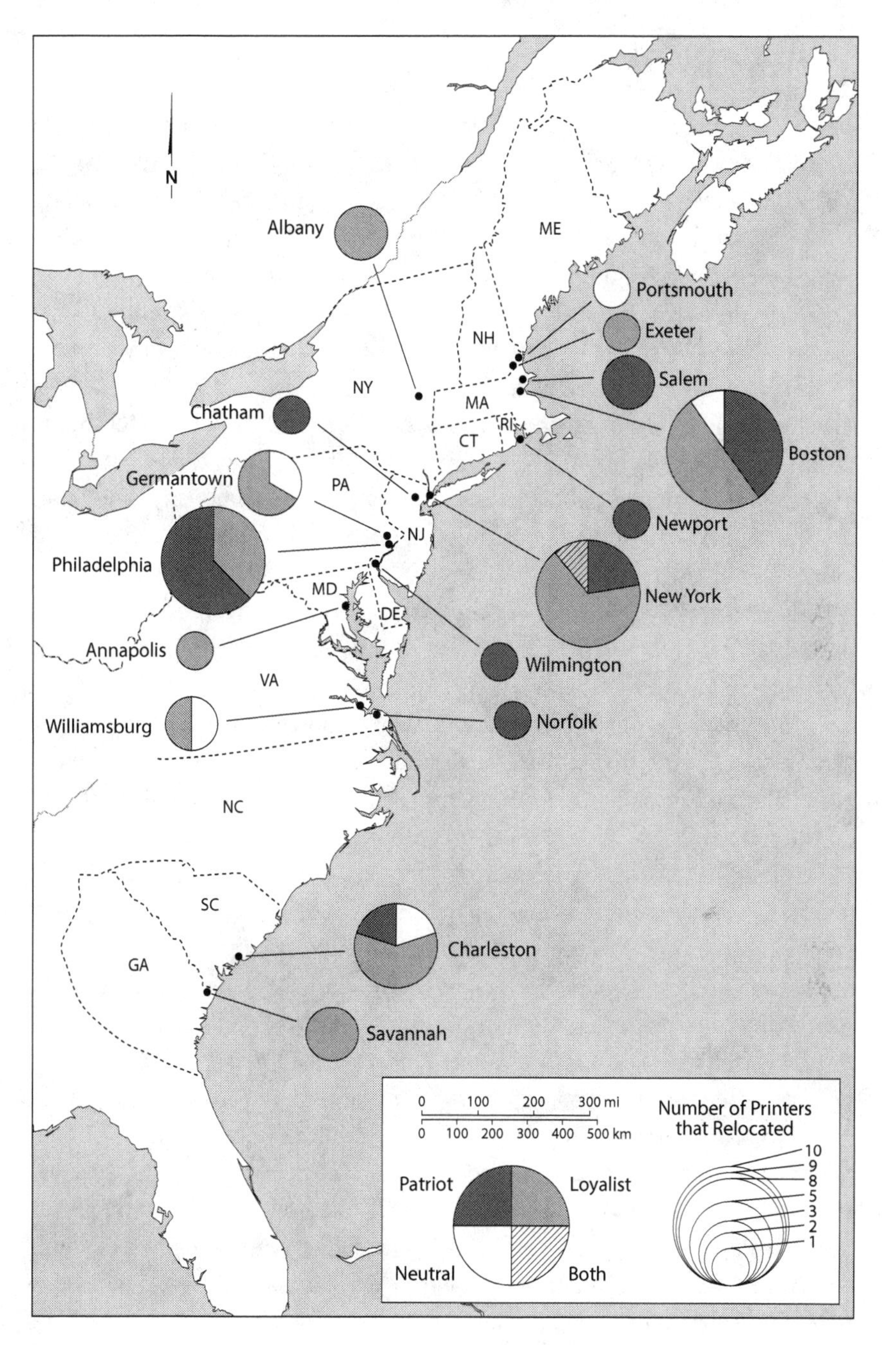

Map 3. Printers relocated during the Revolutionary War, by political affiliation

a few printers attempted to manage the transition between Continental and British control with only brief departures from their offices. These printers, who often suffered ridicule and criticism for their decision, either wavered in their political convictions or lacked particularly strong feelings. Hugh Gaine, for example, fled New York during the British invasion and began publishing his *New-York Gazette* from Newark, New Jersey. The British Army, meanwhile, confiscated his press and began publishing its own version of the *Gazette,* still under Gaine's imprint, from his office. Within two months, Gaine was back in New York at his own press and pledging allegiance to the British. His decision may very well have sprung from a cold-blooded assessment of the Americans' odds, but he nonetheless sought to stay in business by choosing a side he thought might win.[84] Similarly, Benjamin Towne, a Philadelphia printer who continued publishing the *Pennsylvania Evening Post* throughout the war, remained in that city during the British occupation of 1777–78 and then when the British left.

While Americans faced disruptions and difficulties in information flow, the British Army struggled to meet its communications needs. In most cases, Loyalist printers were eager to enter the relative safety offered within British lines, but they tended to move with the army, and therefore had to reestablish offices anytime the British transferred headquarters.[85] Transporting the materials of a printing office was expensive and onerous for Loyalists already on the run. Ambrose Serle, the secretary to General William Howe, believed the British had to fight back against Patriot arguments. To do that, he attempted to set up a system whereby the British Army would subsidize printers to publish pro-British news. American newspapers, he argued to the Earl of Dartmouth in 1776, had "a more extensive [and] stronger Influence" than nearly any other form of communication in creating "the present Commotion." From his perspective, of course, the American press—which printers portrayed as free in defense of liberty—was an "engine" of propaganda directed by Congress. In order to "restrain the Publication of Falsehood" in the pro-American press, Serle proposed a plan for the British to support printers at a cost of about £700 to £800 per year. He knew whereof he spoke. After the British takeover of New York City, he headed the captured press of Hugh Gaine and continued publication of the *New-York Gazette* under Gaine's name. "Ever since the Press has been under my Direction (from the 30th of September)," he noted, he had "seen sufficient Reason to confirm this Opinion" and had received reports that the newspaper had been well-received.[86]

The British did not explicitly adopt Serle's plan to subsidize printers, but officials typically designated a King's Printer in each city they occupied. Yet the British faced a distinct challenge that Americans did not in their efforts to maintain communications circuits beyond the territory they occupied.[87] Serle's proposal aimed to counteract that deficiency by creating a web of Loyalist printers who could communicate with one another. In envisioning such a system, Serle acknowledged that the Patriot press had shaped public opinion. Furthermore, because of the war, the news networks of most Loyalist printers collapsed. They could not do business freely with anyone outside the British lines and the circulation of information from one side to the other, while not completely stopped, was significantly limited.

Loyalist printers therefore utilized a circumscribed set of contacts for news circulation. For James Rivington, who changed the title of his newspaper to the *Royal Gazette* during the war years, the circulation pathways that had operated before the war were essentially closed off for him, and he became dependent on fellow Loyalists and those newspapers that easily crossed the British lines. Only occasionally and in a delayed fashion was Rivington able to print news from the "rebel papers." While the London newspapers would have come from continuing shipping to New York for the war effort, it is likely that his other news came from two sources. First, he got a significant amount of news from Philadelphia when the British occupied the city, in particular the *Pennsylvania Evening Post*, since the British had direct communication links between the two cities. Rivington also appears to have had some access through the porous military lines to newspapers, in particular the *New-Jersey Gazette*, published in Trenton by Isaac Collins.[88] Compensating as best he could, Rivington frequently published secondhand "extracts from Rebel Papers" and filled the *Royal Gazette* with essays and critiques mocking Americans. Successful adaptation in trying circumstances was key to commercial survival.

During the war, political and military leaders on both sides came to see printers as important political actors in their own right. The imperial crisis made it difficult to sustain that fiction; the war made it impossible. Because of their centrality to information circulation, printers' persons and offices became the targets of attack as combatants on both sides of the political equation not only vented frustration and righteous indignation but played politics by other means to silence undesirable voices.[89] In Norfolk, Virginia, John Hunter Holt, the son of John Holt of New York, faced British troops who confiscated his press at the end of September 1775. His alleged infraction was revealing publicly the name of an officer carrying out Dunmore's offer of freedom for slaves who fought with the British Army.[90] According to a letter published in the *Virginia Gazette*, the soldiers claimed that "they want to print a few papers themselves; that they looked upon the press not to be free, and had a mind to publish something in vindication of their own characters." The letter noted further that the soldiers had not acquired any ink, nor had they captured a compositor to help work the press, but that a printer might be on board one of the British men-of-war in the harbor. The author seemed not to relish the idea of the soldiers printing and appeared pleased that Holt had "luckily made his escape" during the raid.[91]

William Goddard, who made a career out of finding trouble in various cities, came to blows with a faction of Patriots in Baltimore. In February 1777, the *Maryland Journal* published a letter by "Tom Tell-Truth," ironically (or so the Goddards thought) praising the offer of peace by General William Howe in the fall of 1776.[92] At that point, the *Journal* was published under the name of his sister, Mary Katherine. Unamused at even the hint of capitulation, a self-appointed group called the Whig Club nonetheless accosted William Goddard to discover the author of the piece. Ever the defender of the free press, Goddard refused "to suffer the Secrets of his press to be extorted from him in a tumultuous way" and was summarily expelled from Baltimore city and county.[93] After the state assembly granted him a return to Baltimore, the Whig Club continued to harass him, culminating in an assault on his printing office in which Goddard and several of his workmen were injured, and Goddard was taken captive to a nearby tavern. Fearing for her brother's safety, Mary Katherine dispatched a neighbor to the tavern to keep an eye on the proceedings.[94] Goddard was again barred from Baltimore and remained in Annapolis until a reprieve from the governor later in the spring.

Patriots policed acceptable discourse by targeting sympathetic printers, but they also reserved energy to harass and disrupt Loyalist printers when they lacked the protection of the British Army. They therefore turned their attention again to James Rivington in New York. As part of a mission "to disarm Tories" in New York City, a group of Sons of Liberty led by Isaac Sears marched from New Haven, Connecticut, to New York in November 1775 with two goals: to round up Loyalist leaders and "to deprive that Traitor to his Country James Rivington of the means of circulat[ing] pison [poison] in Print."[95] According to later news reports, after a force of about one hundred men set up guards outside Rivington's office, a delegation entered and seized his types—the equipment of highest value—and "destroyed the whole apparatus of the press."[96] Most newspaper reports took it for granted that readers knew of Rivington and his alleged misdeeds. A report in the *Virginia Gazette*, however, speculated that the raid "was intended as retaliation for lord Dunmore's conduct, and others attribute it to an apprehension of his relapsing into his former iniquitous publications."[97] For his part, Sears reported to Connecticut's delegation to the Continental Congress that he hoped that his actions would be "a great means of putting an end to the Tory Faction there, for his press hath been as it were the very life & Soul of it." Anticipating criticism for crossing colony lines to conduct such a raid, Sears also wrote that "it wou'd not otherwise have been done, as there are not Spirited & Leading men enough in N. York to undertake such a Business, or it wou'd have been done long ago."[98]

The attack ended Rivington's career in New York for the time being; he next appeared in colonial newspapers just seven weeks later on a list of Loyalists who had taken passage on a ship to London.[99] He returned a year later and served as King's Printer and publisher of the *Royal Gazette*, devoting considerable ink to the foibles and frailties of the American cause. At the same time, historians now suspect that he served as a double agent, providing information to George Washington—including a Royal Navy signals book that aided the French in their naval campaign against the British in the fall of 1781.[100] Publicly he remained a rhetorical target for Patriots. In 1779, in the first issue of the *United States Magazine*, Hugh Henry Brackenridge published an essay (probably penned by his mentor, John Witherspoon) entitled, "The Humble Representation and Earnest Supplication of James Rivington."[101] In the essay, "Rivington" begged for the mercy of Congress, arguing that he could be no danger to the Americans, even with his publications, because, as he noted, he had "expended and exhausted my whole faculty of that kind in the service of the English. I have tried falshood and misrepresentation in every shape that could be thought of, so that it was like a coat thrice turned, that will not hold a single stitch."[102] Rivington's real-life bombast and his promotion of pro-British and anti-American propaganda in his publications continued to make him an easy foil and focus of Patriot anger at Loyalists.

Other Loyalist printers also faced violence at the hands of Patriots. The brothers James and Alexander Robertson had been printers in Norwich, Connecticut, and Albany, New York, since 1773.[103] During the winter of 1775–76, they printed items in Albany for the British Army campaigning in upstate New York led by Guy Carleton, including a battle account "differing widely from that held out to the public by the Friends to the American Cause." The "occasional Newspaper" also contained letters purportedly from John Adams and John Hancock stating that they intended independence along with "sundry other Pieces to awaken the Jealousy of the Loyalists, and put them on their Guard against the Machinations of their insidious Enemies."[104] Because of his Loyalist leanings, James Robertson was forced to evacuate Albany, leaving behind his paraplegic brother. According to their statement made as part of their claim for compensation after the war, Alexander was arrested, imprisoned in Albany, and left for dead when the jail caught on fire. He saved himself only "by lying on his belly and chewing [on cabbages] to prevent being suffocated."[105]

Such accounts were common during the war as printers' offices, homes, and bodies became sites of conflict. Because they served as mediators for the complicated terrain of politics, printers had long found themselves at the center of controversy. During the colonial period, they had asserted the claim that they operated a free and impartial press. Though not always accepted, that argument influenced how political leaders and others responded to the publication of pieces with which they disagreed. The war changed that response because it cast political choices in stark relief. No longer could printers claim to be "meer mechanics"—not because they were more active behind the scenes (though they were), but rather because the very nature of their publications revealed the side to which each printer had staked his or her future. Some printers found a way to create space for themselves to continue to thrive, not least James Rivington in New York. Others, like William Goddard, struggled to keep up with changing political realities.

For printers throughout the new United States, the politics of printing narrowed after 1775 to one question: Which side were you on? That was obviously of critical importance and yet also counterintuitive, because once a printer answered that question, politics became a much less significant concern than it had been during the imperial crisis. Instead, economic concerns rushed to the fore almost to the exclusion of any other questions. Printers worried about whether anyone would buy their newspapers, pamphlets, and broadsides. They struggled to maintain their connections and contracts, particularly with government entities. They considered whether lines of communication would be open for them to transmit their news and receive news from others. And they scrambled to ensure they had the materials on hand to continue operating. Nonetheless, because printers maintained the ability to reach large audiences of people with political news and arguments, control of and influence over the press continued to be deeply important for partisans on all sides.

The Loyalist Diaspora

During the Revolutionary War, most printers—whether Patriot, Loyalist, neutral, or ambivalent—experienced similar setbacks and opportunities. They were sometimes forced from their homes and offices by the movement of armies. When they remained in place, they felt the wrath of their neighbors for their political opinions or the materials they published from their printing presses. They suffered the economic consequences of a business downturn and struggled to maintain themselves by seeking government contracts. As the war wound down, however, Loyalists faced a far different decision than anyone else: whether to try to return to their prewar homes or evacuate to another British territory. Facing the prospect of trying to revive a business that relied on connections among people hostile to their political views, Loyalist printers decided to leave the United States at a much higher rate than the general Loyalist population.

As the war raged, printers with Loyalist inclinations found their options limited geographically to a much greater extent than others active in the trade. Safe only within the confines of British military lines, Loyalist printers had significantly diminished commercial prospects, in some cases subsisting only on publishing for the local imperial government and the British Army. As the British occupied cities and towns along the coast, Loyalist printers found themselves safe or exposed in turn, and several chose to follow the British Army when it left towns, including Boston, Philadelphia, Savannah, Charleston, Newport, and (after the signing of the Treaty of Paris in 1783) New York City.

Loyalist printers were often placed on proscription lists by state legislatures during the war or immediately thereafter, had their property confiscated and sold off, or simply decided on their own that they could not bear to live in the United States. The beginning of hostilities in April 1775 by itself prompted several printers, in particular Loyalists, to pack up and leave the rebellious colonies. William Aikman, an Annapolis bookseller, fled to Kingston, Jamaica, where he again sold books, worked as a stationer, and served as King's Printer until his death in 1784.[106] Similarly, Robert Luist Fowle was forced to flee after "at length he became so obnoxious to the Usurpers" in his native New Hampshire. Fowle served in the British Army first in Canada and then New York until 1782, when he left for London to plead his case before the Loyalist Claims Commission.[107] Even far from the battle lines, therefore, the war affected the decision making of printers.

For most Loyalist printers, therefore, the departure from their towns was permanent. These printers came from across the former colonies, including Portsmouth, Boston, Albany, New York, Philadelphia, Annapolis, Williamsburg, Charleston, and Savannah. Of the twenty Loyalists who left the United States as a result of the Revolution, the largest group (nine) headed to Canada, while five went to the British Isles, and three to the West Indies. The Loyalists who left the United States provided an extensive documentary record of their wartime travails through petitions to the Loyalist Claims Commission, established by Parliament in 1783 to process applications for reimbursement by American Loyalists who suffered property losses as a result of the war.[108] The records document the experiences of some of the printers during the war (in particular depredations they alleged were committed by Americans) and the financial losses they incurred as a result of the war. For example, the brothers James and Alexander Robertson claimed to have lost over £600, including £311 for their printing office and nearly £78 in wages they owed to two journeymen.[109] Similarly, James Humphreys, a onetime Philadelphia printer, claimed damages of £1,713 to his property, asserting that he had started his *Pennsylvania Ledger* in 1775 "at considble expence and risque," and that he "perseveringly supported and published till November 1776" in favor of the British government.[110] Claims most commonly referenced the materials of their trade, other property, lost wages, and revenue forewent.

The case of Margaret Draper of Boston is among the most revealing claims that Loyalist printers made at the end of the war. A descendant of the Green family of printers in New England, she married Boston printer Richard Draper in 1750.[111] The longtime publisher of the *Boston News-Letter*, in 1763 Draper became the printer to the governor and council, a coveted position that was a mark of his loyalty to the Crown.[112] In fact, they commissioned him to publish his newspaper under the title of *Massachusetts Gazette* jointly with the firm of Green & Russell, publishers of the *Boston Post-Boy*, as an official, government-sanctioned publication.[113] Richard died in 1774, leaving the business to Margaret with assistance from his recently joined partner, John Boyle.[114] Draper and Boyle parted ways within months; in his journal, Boyle noted only that the separation occurred "on account of an unhappy Dispute which arose respecting the Contract entered into between her late Husband and myself."[115] Draper then began working with John Howe, a onetime apprentice of her husband. They continued to publish the *Boston News-Letter* through the British occupation and American siege of Boston after Lexington and Concord and into early 1776. Both evacuated with British forces in March 1776 and left for Halifax.[116] From Halifax, Draper then headed to London with her daughter, where she spent more than a decade attempting to gain compensation from the British government.

The commission gathered a voluminous record on Draper totaling well over one hundred pages—far lengthier than the file of any other printer. The record includes several petitions in Draper's name, detailed accountings of her family's property, and Massachusetts legal records verifying her property losses. She was also clearly well connected among Boston Loyalists; her file includes letters of support from Peter Oliver, once the chief justice of the Massachusetts Superior Court, and General Thomas Gage, commander of British forces in Boston and the military governor of Massachusetts from 1774 to 1776. In her documentation, Draper laid out the sympathetic case of a widow who took up her husband's defense of the Crown under difficult circumstances and the ostracism she faced in her hometown as a result. In her 1784 petition to the commission, Draper outlined how she and her husband had served the Crown and imperial interests from their Boston print shop. Drawing on the long history of the Green and Draper families in printing, Draper argued that their government work "afforded Opportunities to the Ancestors and Husband and to Your Memorialist herself to manifest the most undoubted and eminent Acts of Loyalty to the Crown and Parliament of Great Britain." She noted the work of the *Massachusetts Gazette* to promote the government's view and suggested that she and her husband became "Objects of public Disgust" as a result.[117] Her efforts, of course, were intended to persuade the commissioners to grant reparations and a living stipend to a nearly sixty-year-old widow and her daughter, so Draper underscored and reiterated her family's service to the imperial government of Massachusetts throughout her letters and petitions. As one of the few printers to remain in Boston after April 1775, she lost almost all support, and all but had to evacuate Boston when the British Army left for Halifax because she had "rendered herself more the Object of the rancor and Malice of the Rebels than most others."[118]

Because the Claims Commission was primarily interested in financial losses, the commission records provide a fascinating and detailed window into how these Loyalist print shops operated and the networks they maintained not only before and during the war but afterward as well. In Margaret

Draper's case, she at least claimed to have been operating what would have been a relatively significant printing office. In a schedule she prepared in 1784, for instance, she valued her printing office at approximately £130 and calculated that the *Massachusetts Gazette*, included debts owed, was worth more than £500.[119] But these calculations could not be done only from England. Claimants therefore required assistance from American governments to determine the value of their property in the United States. In Draper's case, that included an additional assessment of her printing office in 1784 from Massachusetts authorities, which valued her printing materials at £190 and estimated her annual income at £520. Because the state sought expert judgment, the men who conducted the review turned out to be some of Draper's old nemeses: Patriot printers Thomas Fleet, John Gill, and Benjamin Edes, with the entire valuation certified by John Hancock.[120] Draper ultimately spent more than a decade trying to gain restitution—the letters in her file date from 1777 to 1789. In the end, she received for most of the time a living stipend of £60 per year, which was briefly increased to £100 during the 1780s. We know little of her life after that date except that she died in London in early 1807.

Evacuation also split some printing families where members did not share the same politics. In Pennsylvania, the Sower family (sometimes spelled Sauer) had been central to the German-language press that developed in that colony for three generations (all of them named Christopher). During the 1760s and 1770s, Christopher Sower Jr. and then his sons Christopher Sower 3rd and Peter Sower published the *Germantowner Zeitung*, which continued publication into 1777. At that point, the military campaign arrived in Pennsylvania. Christopher Sower Jr. remained neutral and refused to take an oath of allegiance to the United States or to join the military, in large part because he was a member of a German Baptist group known as the Dunkers. He sought refuge in Philadelphia during the British occupation, during which time Patriots destroyed his Germantown printing office. His sons, Christopher 3rd and Peter, were more actively Loyalist. They both fled to New York City in 1778, where Christopher took up work as a bookbinder. At the end of the war the younger Christopher left for England, and Peter headed for the New Brunswick town of St. John, where he studied medicine. Christopher 3rd returned to North America in 1785, joining his brother in St. John and becoming publisher of the English-language *Royal Gazette*. After serving as the St. John postmaster, Christopher 3rd returned to the United States to live with his son and died in Baltimore.[121]

For those Loyalists who evacuated with the British to other colonial possessions or England, the story of their contribution to and participation in these news networks and communications circuits ended at the same time as the war. Though a few, most notably Margaret Draper, had deep ties to colonial printing, many had migrated to colonial North America within a few years before the war began. They all faced the same depredations and difficulties, but without ties holding them to the United States, there seemed little reason to stay. They spent the rest of their careers as British subjects in the Bahamas, Jamaica, Nova Scotia, and New Brunswick and in England. They sought compensation from the Claims Commission and, like so many other Loyalists, were frustrated that they did not receive what they thought they deserved. A few Loyalists did remain in the United States, for reasons that will become clear in the next chapter. But the war ended any debate in the United States about whether to proclaim allegiance as American or Briton.

The United States and Britain agreed to the Treaty of Paris in September 1783, formally ending the war and bringing recognition to the independence of the United States. As negotiators wrapped up in Europe, Americans were beginning to put the pieces back together, and Loyalists were making their way to other British territories. For the printing trade, the war had been devastating. It cut off standard pathways of communication and supply, making it vastly more difficult for printers to gather the intelligence and news required to publish their newspapers, almanacs, and other printed matter, not to mention the paper and ink with which they were printed. Printers became a focal point for rhetorical attack and physical violence. The war scattered printers from their seaport offices and homes. With neither the option nor the desire to return to what would amount to political exile, many Loyalists began to forge new lives in other parts of the British Empire. The Revolutionary War thus defined the development of the printing trade (as so much else) into the early United States. The eight-year conflict reshaped the personnel of the trade, sweeping out many of those loyal to the Crown and dispersing others. At the same time, printers adapted to the circumstances as best they could, seeking out opportunities with governments, armies, and wherever else they could. The printing trade defined the Revolution as the presses of printers throughout North America vied to depict the struggle on their terms. In so doing, they set the stage for a new national communications infrastructure that would take hold in the 1780s.

Rebuilding Print Networks for the New Nation

Matthias Bartgis was ready to start a newspaper. Born in central Pennsylvania to German immigrant parents, Bartgis served his apprenticeship with William Bradford, one of the leading printers of colonial Philadelphia. During the Revolutionary War, Bartgis set out on his own in the rural town of Frederick, Maryland, about fifty miles west of Baltimore. For the first several years, his printing office produced small one-off printing jobs as well as almanacs in both English and German. But by the end of 1785 Bartgis was prepared for a bigger challenge. He proposed printing the *Maryland Chronicle* to bring news and information to the interior towns that ran from Frederick to the south and west into Virginia along the ridge of the Appalachian Mountains. At the same time, according to one printing historian, he also proposed to print a German-language paper, but no evidence of the paper survives.

In his opening address to readers, dated January 4, 1786, Bartgis extolled the virtues of his trade. "Amongst all the discoveries which human sagacity has made," he wrote, "one of the most important and useful to mankind is PRINTING." The greatest figures in history had sought to get their ideas into print and "none but sordid minds . . . ever wished to extinguish this light of Heaven." Furthermore, he contended, the press was crucial to the proper function of a free society. In the hands of "tyrants," the press "conveys nothing to the people but a stream of poison, which clouds the rays of light, puts out the public eye, quenches the fire of patriotism, and excludes every celestial beam from the human mind."[1] As the first newspaper in that region of the interior, Bartgis also sought to educate his readers on how to subscribe to a newspaper. Knowing that his rural customers were spread thinly across the Maryland and Virginia hills, Bartgis explained a system he suggested was the "practice of the Eastern States," in which fifty-two residents of a town jointly subscribed to a newspaper for the year, each taking responsibility for a week in order to share costs. Finally, to give his newspaper a mark of respectability, he crafted his own Latin motto: "Qui nova desiderat, pervolvat viscera nostra: affero delicias, fortuitosque casus," which translates as "Let whoever longs for the news turn through my contents: I report delightful, as well as, fortuitous events."[2]

By the end of his career, Bartgis opened additional printing offices in York, Pennsylvania, and Winchester, Virginia, both in the backcountry of the new United States.[3] His story exemplifies one of two directions the printing trade developed in the 1780s. He trained with a colonial master but did not become active until during and after the Revolutionary War. When he sought out an opportunity to open his own printing office, he found it in the continental interior, moving west with many of his fellow Americans of European descent. In these small rural outposts, many printers took up the practices that had sustained the businesses of their colonial forebears. They found it difficult to drum up enough business to continue operating, so they appealed to all comers rather than take a partisan stance in their publications. Printers commenced newspapers at a startling clip in this period but published them only weekly, focusing on the distribution of commercial and political news both within the United States and across the Atlantic. And their offices continued to serve the general interests of customers by publishing almanacs, blank forms, and other custom printing jobs.

In the burgeoning Atlantic coast cities, by contrast, the printing trade was becoming more diversified and competitive as printers took advantage of expanding audiences for their publications. Printers in New York, Philadelphia, and Boston faced even more pressure to keep up after the war than during the early 1770s. Older printers returning from exile expected to take up the same roles they had held a decade earlier, and new entrants into the trade sought to take official business away from them in order to make their mark. Because the trade continued to expand along with the port cities, printers sought new opportunities to succeed, whether by publishing newspapers more frequently—including the first daily newspapers in the United States—or by advancing their business into the world of book publishing, previously untouched in North America.

To reconstruct their trade in the aftermath of the Revolutionary War, printers had to overcome the strong headwinds of a struggling economy as well as the continuing difficulties of operating a printing office that could sustain a family. The 1780s also saw the beginnings of a bifurcation in the trade. Printers in rural areas, particularly those newly settled in the American interior, found limited options in a region that was, for all intents and purposes, colonial. In growing cities, by contrast, printers sought to innovate to reach a growing audience hungry for news of both American and European politics and commerce. These trends were exacerbated by conflicts over the legacy of the Revolution. Communities debated whether Loyalists who chose to stay should be reintegrated into society, and those who had supported the American cause fought over who could lay rightful claim as the most patriotic. At the same time, all printers continued to assert the ideology of a free and open press as the "bulwark of liberty," portraying their work as central to the new project of creating an American republic. These trends reached a crescendo in 1787 and 1788 as printers, among the most broadly networked tradesmen in the United States, put their publications firmly on the side of ratifying the newly drafted Constitution.

The Shifting Economics and Demography of the Printing Trade

As the Revolutionary War came to a close, the United States attempted to manage both a renewed demographic and geographic expansion while navigating harrowing economic straits. With the departure of the British Army (for the most part) from the trans-Appalachian West, white Americans began to flood across the mountains in the 1780s in search of more land and economic opportunity. Land that would eventually become Ohio, Kentucky, and Tennessee became popular destinations for these settlers. In addition, the population of western parts of the thirteen states mushroomed, as did other interior areas, most notably Vermont and Maine. No longer hampered by imperial policies that attempted to curtail such settlement, Americans poured West to displace Native populations and establish European-style towns and villages.[4]

At the same time, the United States faced a massive postwar economic crisis. From the beginning of the Revolutionary War in 1775, and accelerating after the conclusion of hostilities, the American economy suffered from debilitating inflation and a serious deficiency of specie. At the national level, Congress struggled to finance the war effort, in no small part because of the strictures that the Articles of Confederation placed on its taxing capabilities.[5] Robert Morris, the superintendent of finance during the early 1780s, promoted a general impost in 1782 to raise funds, but it failed when most states simply refused to provide the requisitioned funds. Congress experienced the same issue again a few years later when faced with particularly strong recalcitrance from New York, the nation's burgeoning commercial center.[6]

Meanwhile, the burden of debt wreaked havoc for individuals, especially those whose Continental Army service terminated with nothing in payment other than largely worthless loan certificates. Across the country, ordinary farmers and businessmen found themselves beset with offers from speculators to scoop up Continental certificates at a fraction of face value, judgments for debt, and ever-increasing prices.[7] These two countervailing trends influenced printers as much as anyone else. They traveled west, set up shop in new towns, and used whatever contacts and connections they had to integrate these settlements into the American communications infrastructure. Printing thus remained a precarious commercial venture, especially outside the context of the Atlantic ports, the main centers of American commercial activity.

These factors notwithstanding, the printing trade grew by leaps and bounds through the 1780s. During the Revolutionary War, the numbers of printers in the trade and of newspapers published both plateaued as commercial opportunities vanished and the ability to transmit information was severely curtailed. In the last full year of the war (1782), 113 printers were active in the United States; by 1789, when the federal government began operations, 229 printers were active—the number nearly doubling in just seven years (fig. 11). As seen in fig. 9, newspapers appeared even more quickly, rising in number from 45 in 1782 to 102 in 1789, representing growth of 127 percent. Boston, New York, and Philadelphia, the three towns where most printing took place in the decades before the American Revolution, remained dominant. However, the trade spread so much and so quickly that printers in these areas gradually represented a smaller proportion of the total number of printers. Between 1782 and 1789, offices appeared in thirty-eight new towns, raising the total from thirty-nine at the end of the war to seventy-three when the federal government began its work.[8] These included new inland state capitals such as Harrisburg, Pennsylvania, and Augusta, Georgia, western settlements such as Pittsburgh, Pennsylvania, and Lexington, Kentucky, and additional offices in heavily settled states such as Petersburg, Virginia, and Hudson and Lansingburgh, New York. That brought down the proportion of printers in the three biggest cities from 45 percent in 1782 to 38 percent in 1789.

The massive expansion of the trade away from the Atlantic coastline led to a bifurcation in the business and political strategies of the men and women engaged in printing. In more rural areas, especially those where printing was appearing for the first time, the colonial model of the trade persisted as a way to ensure broad-based support in the community. In the growing urban

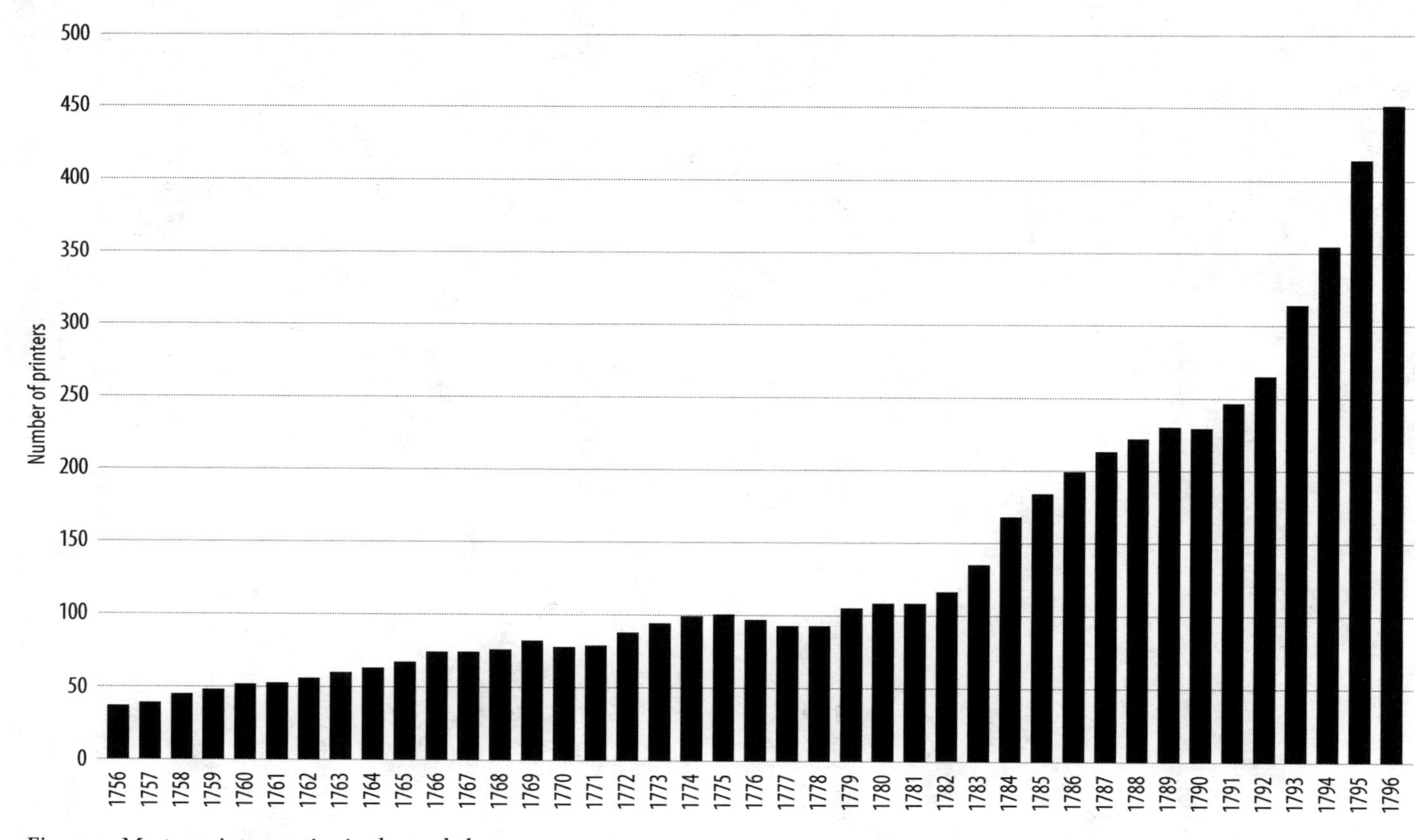

Figure 11. Master printers active in the trade by year, 1756–96

centers, by contrast, innovation in commercial practices became necessary as printers competed for business.

The Persistence of the Colonial Model

After the war, tens of thousands of white Americans headed west to areas along and across the Appalachian Mountains, settling on land that the British had once decreed was off-limits for colonial settlement. These regions were in many ways the first colonies of the United States. Whether on land claimed by a state or the national government, these areas faced issues of sovereignty and authority, paramount among them their relationship with an imperial center, whether that be a state capital such as Richmond or the shifting national capital.[9] For those men and women who set up as the first printers in their new communities, the trade thus looked remarkably similar to its status a half century earlier, as if the political upheavals of the 1760s and 1770s had never happened. Rural offices offered services broadly to the public, printed weekly newspapers that rarely turned a profit, and continued to project the image of a printer as public servant rather than opinionated political participant. Geographic expansion, in other words, brought both opportunities and challenges. On the one hand, every new village and town in the American interior wanted a printer (and the newspaper that often accompanied him or her) to demonstrate its permanence, but most could not offer sufficient business—a situation akin to what many printers faced in the British colonies before 1763.[10] At the same time, many of these settlers arrived in their new towns with little experience in the printing trade. In fact, it has been hard to confirm for many of the printers active in the 1780s whether they had served formal apprenticeships or had any background or training in the printing trade. Furthermore, in many cases their careers proved to be ephemeral. The retention rate for trade personnel returned to equilibrium in the 1780s but settled at a lower rate: on average, about three-quarters of printers active during a given year in the 1780s were still in the trade five years later (fig. 12). The rate was, it should be noted, considerably lower than in the late colonial period (when it hovered above 80 percent over a five-year period).

Most importantly, printers in colonial areas continued to depend on networks and connections. Printing as a trade required a critical mass of readers and consumers in a local market (even as it made claims to universality through its reliance on the circulation of distant news). Inland communities often had small readership and advertising bases on which to build, which could make it difficult for a printer to make ends meet. Vermont offers a useful example

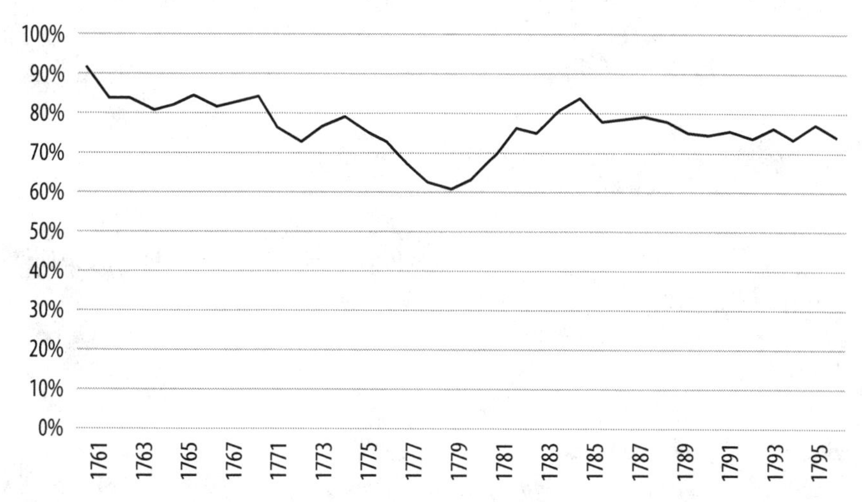

Figure 12. Proportion of printers active in a year who remained in the trade five years later

of how printers and government officials worked together to bring printing to newly settled territories. Formed from territory claimed by both New York and New Hampshire, Vermont exploded in population after the Revolutionary War with in-migrants traveling up the Connecticut River from the south and continuing to enter from both New York to the west and New Hampshire to the east.[11] Because both older states claimed the territory, Vermont existed in a geopolitical limbo between its founding in 1777 as a republic and its admission into the Union in 1791 as the fourteenth state. State leaders nevertheless made proactive efforts to extend lines of communication both among the communities within the state and between Vermont, New York, and New England.

The sudden rise in population created new demand among Vermont officials for printers as they looked to develop connections with the rest of the United States. Officials eventually attracted printing offices to Bennington and Windsor, two of the state's major towns. Anthony Haswell came to Bennington in 1783 to take up a position as postmaster for the state. He had migrated from England in the late 1760s with his father and brother and was initially apprenticed as a potter. But after his father and brother returned to England, the Boston Overseers of the Poor indentured Haswell to work with Isaiah Thomas in 1771.[12] After working for Thomas in Boston and Worcester and for a time publishing Thomas's *Massachusetts Spy* on his own, Haswell was ready for new challenges and shifted his operation north, where he published the *Vermont Gazette* and began printing for the state government.

The opportunities in Vermont also drew the attention of the Green family network of printers in the form of Green in-laws Alden and Judah Padock Spooner. Like Haswell, they followed opportunities north into the upper Connecticut River Valley after the Revolutionary War. Born and raised in New London, Alden and Judah trained in printing with Timothy Green, who was married to their sister Rebecca. Green then set up Judah with materials for a printing office in Norwich, about fifteen miles up the Thames River. According to Judah's son, however, Green prevented Spooner from starting a newspaper that might compete with his *New London Gazette*. When fighting began in 1775, Judah rushed to Boston, where he served and was wounded at the Battle of Bunker Hill. After the war, Green again came to the Spooners' aid by providing them materials to set up a printing office at Dresden, Vermont (now Hanover, New Hampshire, home to Dartmouth College). There they began what likely was Vermont's first newspaper, the *Dresden Mercury*, in 1779. Unfortunately, Judah suffered greatly from his war wounds, according to his son, and abandoned his family for a decade to pursue phantom legal action.

In his brother's absence, Alden Spooner moved his printing office to Windsor, Vermont, which served as the state's capital. He also took on a new partner, George Hough, who had been an apprentice in Connecticut at the same time as the Spooner brothers. Hough, however, had served his apprenticeship with Loyalists James Robertson and John Trumbull, publishers of the *Norwich Packet*—a newspaper that had threatened Timothy Green's business in New London in the early 1770s. The two commenced the *Vermont Journal, and the Universal Advertiser* in August 1783. As with other new newspapers, Hough and Spooner declaimed in an opening address to customers that a newspaper was always "of public utility" and that the absence of a paper in Vermont (Haswell's *Vermont Gazette* notwithstanding) meant that "many disadvantages have accrued to the people in consequence thereof." As a rural paper, Hough and Spooner emphasized that they would "make it their peculiar study to keep the press FREE, uninfluenced by *party* or *prejudice*."[13] Like Haswell, Hough and Spooner printed for the state government.[14] The issue of news circulation was also at the forefront of concerns for both new offices. The Continental Post Office did not extend northward in interior New England past Springfield, so the printers began developing post routes of their own. They also lobbied the state legislature for the establishment of post routes and offices. In 1784 the legislature obliged, setting five offices within the state, creating a route to Albany to link Vermont's post with the rest of the United States, and appointing Haswell as the state's postmaster general.[15]

Just as it was a point of civic pride to have attracted a printer and newspaper to a community, it was quite the black eye to lose one. For instance, Isaiah Thomas advertised in April 1781 that he needed to get his subscriber base up to 750 (and get most of them to pay, as printers commonly complained) in order to make the *Massachusetts Spy* a sustainable enterprise in Worcester.[16] A week later, an anonymous author named "Monitor" wrote an essay for the *Spy* to attest that the newspaper was "*of great public utility, and of much advantage to individuals.*" He framed his argument in terms of international affairs, noting that the "present day is pregnant with mighty events!—events unparalleled in the annals of ancient or modern history." Citing the entry of the Dutch into the Revolutionary War on the side of the Americans and the intrigue of the Bourbons in France and Spain, Monitor lamented, "Those people who are so indifferent to objects of their welfare as to exist in a state of insensibility of the great things working out for their happiness," that is, those residents of Worcester County who had not subscribed to the *Spy*, "scarcely merit the blessing resulting from the happy effects of being freed from the *clanking chains* that despotism and tyranny were forging for them and their posterity!" Monitor was especially concerned that his fellow citizens could not see the consequences of their indifference. "If the PRESS is suffered to go away," he intoned, "the more interior stations of the county will be in darkness as to the ideas of publick affairs."[17]

The need for printing in unprofitable locations forced some to consider what once would have been unthinkable: welcoming back Loyalists they had condemned during the war. In Georgia, the mutual needs of printer and government, as well as the demands of the printing trade, eventually converged in favor of reconciliation. James Johnston was the first printer in Savannah, Georgia, and had published the *Georgia Gazette* since 1763. He abandoned the city when the local Committee of Safety began investigating his commitment to the Patriot cause in January 1776. For the next several years, Georgia struggled to find a printer who could manage the printing of the government, including the laws of the state, because overall demand for printing work was so low. Johnston returned to Savannah during its British occupation from 1779 to 1782, serving as King's Printer and publishing his newspaper as the *Royal Georgia Gazette*. This earned him the enmity of the Patriots controlling the state government. In 1782 the restored state government placed him on a list of 117 Loyalists banished forever from the state and began the process of confiscating and selling his property. Within a few months, however, tempers cooled, and the assembly rescinded its earlier decision, in part because it faced a shortage of printers. Within a few days of passing an act allowing his return in January 1783, Johnston resumed work as state printer.[18]

Nonetheless, Georgia remained a backwater of both information and commerce. When Johnston rekindled his newspaper as *The Gazette of the State of Georgia* in early 1783, he did so with almost no advertising, which meant that he had to rely on government patronage just to put out the newspaper each week. Advertising in the *Gazette* did not begin to return in full force until the fall of 1783. This is not surprising, given that the city had fallen under British occupation for three years, but it nonetheless hampered Johnston's independent business prospects. No more helpful was the flow of news into Savannah, which came in fits and starts with arriving ships rather than the steady flow that most cities from Williamsburg north had achieved by that time. In fact, during the early part of 1783 most of the news that Johnston published from outside Savannah came by way of St. Augustine in East Florida, which remained a British possession and had become a center for Loyalist refugees from the Lower South. Whatever Johnston's political sympathies at that point, he continued to rely on information circuits populated by his former Loyalist colleagues and likely suffered from Georgia's relative isolation. By the mid-1780s Johnston had reintegrated himself as he readily took up the task of publishing resolutions of the Continental Congress and of the new state government, while running advertisements and selling pamphlets for the Georgia chapter of the Society of the Cincinnati.[19] Stories of a similar cast played out in several dozen other communities as Americans evaluated how to establish print and communications in a national rather than imperial context.

Growth and Competition in Urban Seaports

While printers struggled to scrape together business in the countryside, their counterparts in the burgeoning cities of the Atlantic coast faced far different circumstances. Each of the major publishing centers of colonial America—Boston, New York, and Philadelphia—faced an uncertain outcome at the end of the Revolutionary War. Each city had been occupied during the war by the British Army. Boston, where the war had begun, had been all but completely free of British influence since early in 1776 but nonetheless faced economic travails. Philadelphia was a focal point of the fighting on both sides for several years in the late 1770s. New York was the last city ceded by the British, who did not leave until November 1783 after news of the Treaty of Paris had reached North America. The economic downturn that plagued the United States in the 1780s certainly had an impact on these cities, but at the same time they began a process of rapid expansion in both raw numbers and the complexity of their economic activity. For printers, that process led to a contest to claim the legacy of the American Revolution, new competition for business, and opportunities to expand or innovate with new practices. These conflicts often manifested across generational lines as older printers struggled to stay in business and remain relevant within their communities against a rush of new, younger entrants into the trade.

As the three biggest cities in the United States mushroomed during the 1780s, printers began to catch up with trade practices already common in major European cities. One such area was in the publication of daily newspapers. Before the Revolution, nearly every American newspaper published on a weekly schedule. A few attempted to publish two or three times a week, but none carried on that schedule for very long. After the war, however, dailies caught on in Philadelphia, New York, and Boston. Benjamin Towne started the first daily newspaper in the United States in 1783, the *Pennsylvania Evening Post, and Daily Advertiser*, which signaled increasing competition and capacity in the major coastal cities.[20] Within a few years, several more dailies appeared in the three major seaports, including newspapers that increased their publication frequency, such as the *Independent Gazetteer* (Eleazer Oswald) and *Pennsylvania Packet* (Dunlap and Claypoole) in Philadelphia, and new newspapers, such as the *Daily Advertiser* (Francis Childs) in New York. In these early years, the daily newspapers served a different function from the weekly newspapers to which most Americans were accustomed. Rather than including an eclectic mix of commercial, political, social, and cultural news from around the United States and the British Atlantic world, daily papers focused more narrowly on commercial news: ships entering or clearing customs, the prices of major trading goods such as tobacco, sugar, and other commodities, and—as always—advertising.

Innovation was an important but small part of the picture; more widespread was generational turnover, which reshaped the trade in major cities. By the end of the Revolution, many of the printers who had been most active during the imperial crisis and early in the war years were entering the latter years of their careers and lives. Men who had helped create Sons of Liberty organizations in their towns—Benjamin Edes in Boston, John Holt in New

York, William Bradford in Philadelphia, Alexander Purdie in Williamsburg, and Peter Timothy in Charleston—had passed the age of fifty, were retired, or (in the case of Purdie and Timothy) died during the war. The prominent Loyalist printers who remained in New York after 1783, Hugh Gaine and James Rivington, were both in their late fifties. Though age could provide these men and women with credibility in their communities and the advantage of well-developed networks, printing remained at heart brutal physical labor that wore down the bodies of its participants. As age took its inevitable toll, these printers faced an additional threat from young, ambitious, and more energetic printers arriving on the scene. The rising generation of young printers attempted to displace older printers both from their positions in commercial circles in cities and from official posts, which remained more reliably lucrative. Printers of the older generation thought they had earned the right to those posts for their devotion to the Revolutionary cause and publicly campaigned to retain or earn positions as printers to the state government by appealing to the patriotism of the populace and legislators.[21] Those pleas, however, often did little to attract customers in the face of younger and more energetic competitors.

The end of John Holt's career is both enlightening and tragic in this regard. Holt returned to New York City from Kingston on the heels of the British departure in 1783. In the first issue of his newspaper published in New York in seven years, Holt recounted his service as a printer during the imperial crisis. "*When the Differences between Great-Britain and America first arose*," Holt wrote, he "*laboured as far as the influence his Business gave him, extended to remove the Cause*." After the battles of Lexington and Concord, when "*a Reconciliation became impracticable*," his newspaper, the *New-York Journal*, and his other publications "*tended to animate his Countrymen to a vigorous Defence of their just Rights and Freedom.*" His own efforts, though but "*a drop to the Ocean*," when aggregated with the actions of others led to the triumph of the Revolution.[22] Over the next two months, Holt continued to invoke his service to the state and country in the Revolution as justification for people to subscribe to his newspaper. In January 1784 Holt changed the name of the newspaper from the *Independent New-York Gazette* to the *Independent Gazette; or the New-York Journal Revived.* He used the opportunity to address his readers directly again, noting that "the *public good*" had always been "the principal object of the Printer's pursuit" and again detailing the loss of property and business he had suffered from seven years in exile.[23] He also pointed out, perhaps in a bid for sympathy, that he had returned to New York "in an exhausted state"—and indeed he had, for he died barely three weeks after publishing that notice.[24]

His death set off a scramble for the government contract that he had hoped to recoup. His printing office passed to his wife, Elizabeth Holt (who was the daughter of printer William Hunter). She pledged to operate the press and publish the newspaper in the same manner as her husband, making the same appeal to Revolutionary zeal: "As his press, conducted with superior talents and ability, was consistently devoted and directed to the support of liberty, virtue, and independence, so it shall still remain sacred in the same exalted cause."[25] A month later, she made several changes to the newspaper that staked her claim to commercial legitimacy and prominence. First, she restored the newspaper's prewar name—the *New-York Journal*—and added the words "State Gazette." In addition, she appended a large image of the New York State seal atop the masthead. These new marks emphasized her position as state printer and her access thereby to the upper echelons of New York's political life. She also reiterated her husband's appeals to patriotism as a justification for subscribing to the *Journal*. In addition to restoring the name, she resumed the number of the prewar journal, counting the first paper with the old name as number 1950.

In her address to her customers, she reminded her readers of her family's contributions to the patriotic cause in the face of enormous opposition from others: "The generous and enlightened part of our community," she wrote, "have long since manifested their entire approbation and satisfaction of the doctrines and sentiments it contained;—while they beheld with an eye of indignation and contempt, the *partial* and *execrable* publications, set on foot, supported, and published under the influence of an abandoned *British* Administration, and dealt out and distributed by their more infamous *hirelings* and *tools*, their royal associates in every species of villainy and falshood."[26] She trusted her audience to understand that she referred to Hugh Gaine and James Rivington, still active in the city. As Loyalists, she suggested, they did not deserve the business of any New Yorker. These arguments worked, at least for the time being; the New York State Legislature continued to employ her as state printer, and she operated the printing office. But the victory was short-lived: within a few years Holt passed the printing business to Eleazer Oswald, her son-in-law, and lost her state printing contract.

Appeals to patriotic service thus had limitations, which proved true for the most zealous printer in the most rebellious city. Benjamin Edes was likely the most important anti-imperial printer in the colonies. He was central to the activities of the Sons of Liberty, worked closely with the Boston Committee of Correspondence, and hosted Samuel Adams, John Adams, John Hancock, and other Patriot leaders in his office to write anti-imperial pieces for the *Boston Gazette*. But the war ruined him. The *Boston Gazette* ended its run in 1775 when Patriots evacuated Boston, as did the partnership Edes had enjoyed for twenty years with John Gill. He returned to Boston after the war and resumed printing, but he never regained the audience he had once reached. His political principles seemed out-of-touch and out-of-date in the postwar period, and business difficulties left him destitute when he died in 1803.[27]

New York's remaining Loyalist printers faced an even greater hurdle because of their apostasy. As the last city occupied by the British Army in North America, it housed by the autumn of 1783 not only a long-standing population of Loyalist locals but also a large population of Loyalist refugees from around the new United States, most of whom were using the city as a way station en route to other sites in the British Empire. Whether Loyalists would remain in the city after the British departure—and, furthermore, whether returning Patriots would *permit* them to do so—prompted a heated debate. Many, for example, had been placed on proscription lists by the New York State legislature during the war, which intended to bar them from remaining in American territory and to confiscate their American property.[28] John Holt, embittered about his experience in exile, repeatedly threatened Loyalists after his printing office in New York City reopened. In mid-December, for instance, he published a "card" warning Loyalists that they were not experiencing the calm after the storm but would instead find that "the calm which the enemies of *Columbia* at present enjoy, will ere long be succeed by a bitter and *neck-breaking* hurricane."[29] The very next week, "An Observer" suggested that upcoming Assembly elections would bring into office men who would "strike terror to those that have been base turncoats from the cause of Liberty."[30] By early January, his newspaper was circulating the suggestion that Loyalists be banned from New York for seven years.[31] For many Patriots, this was a typical reaction, as anger at the remaining Loyalists grew as they refused to leave and asserted their right to stay.

Both Loyalist printers in New York—James Rivington and Hugh Gaine—did precisely that. It should not be surprising that Gaine sought to stay, not least because he had proved to be politically fickle during the Revolutionary War. When the British first took New York, Gaine fled across the Hudson River, but within a few months, with the American war effort looking bleaker, he returned to the city, took an oath of loyalty to the king, and resumed his printing business on the island of Manhattan.[32] As the British departure drew near, Gaine reduced the scale of his business, ending the run of his newspaper, the *New-York Gazette, and Weekly Mercury*. Instead, he sought to focus on producing books and other printing materials. He nonetheless still faced the censure of New York audiences. Early in 1783, for example, a satirical poem appeared in several newspapers mocking Gaine's supposed petition to the New York State Senate (no original has ever been found). Probably written by Philip Freneau, the poem excoriated Gaine as a venal and immoral businessman with wayward political leanings.[33] Before the war, the fictional Gaine wrote, "Yes, I was a whig, and a whig from my heart,/But still was unwilling with Britain to part—."[34] His conduct during the war was equally portrayed as self-serving. The poem recounted that he fled from New York to Newark but described his experience as unsatisfying. By the end of the poem, the fictional Gaine once again pledged his allegiance to the United States and offered his services to the Americans—though the poem makes clear that his allegiance is available to any ruling party:

> My press, that has call'd you (as tyranny drove her)
> Rogues, rebels, and rascals, a thousand times over,
> Shall be at your service by day and by night,
> To publish whate'er you think proper to write;
> Those *types* which have rais'd George the third to a level
> With angels---shall prove him as black as the devil,
> To him that contriv'd him a shame and disgrace,
> Nor blest with one virtue to honour his race![35]

To Patriots, Gaine's strategy belied his allegiance only to his commercial interest and his complete lack of a moral compass. His critics unironically positioned themselves as uninterested in commerce in order to drive Gaine out of business. It did not have its desired effect. The poem, which lambasted and lampooned the aging printer, may have embarrassed Gaine, but it certainly did little to motivate him to leave New York. Gaine faced some vitriol after November 1783 but was largely successful upon his return to the book trade. At the end of the 1780s, he was conducting business with Isaiah Thomas, one of the leading Patriots during the Revolution, exchanging books between their respective operations.[36] And in early 1789, Gaine led a movement of New York printers attempting to publish an American edition of the Bible. The publishers argued in a printed circular letter, "It must be a Matter of sincere Regret to every Well-wisher to *American* Manufactures, that now being an independent Nation, we must have Recourse to another Country for that very Book (viz. the Holy Bible) the printing and publishing of which, dignifies every Christian Country where it is manufactured; and even the wisest Legislatures have given Sanction and Encouragement to particular Bodies, to have correct Editions of the same, neatly printed."[37] An American nation, in other words, deserved an American Bible.

James Rivington, who spent most of the war years pugnaciously challenging the American cause, also managed the transition back into American society. As the war wound down, he gradually moderated the tenor of the *Royal Gazette*. The *Gazette*, in fact, is a useful mechanism through which to understand the shifting terrain of Loyalist thinking as the end of the war, and British recognition of independence became inevitable. Most fighting ended in 1781, so through 1782 and early 1783 Rivington began to carry more news about peace negotiations from London and Paris. The *Royal Gazette* saw a concurrent decline in references to American newspapers as "Rebel Papers," which had been a common moniker during the war. Rivington also began to face the reality of Loyalist diaspora. He printed letters through the summer of 1783 from settlers in Nova Scotia and New Brunswick about the delights of the Canadian Maritimes.[38] In June 1783 he offered his readers a "Description of the Bahama Islands, taken from the Political Essays," as material that "may not be disagreeable to those who incline to settle there."[39] He also chronicled the formation of what became the Loyalist Claims Commission, publishing not only the act creating it but also several announcements pertaining to its early activities.[40]

Rivington himself showed no signs of leaving New York. He remained a committed businessman and adhered to many of the customs of the trade. This included publishing the advertisements and announcements for the return of Patriot printers from exile and the start of publication of their newspapers.[41] Rivington clearly signaled his intent to continue publishing in the city: he advertised for new pressmen, changed the name of the newspaper to *Rivington's New-York Gazette*, and removed the royal arms from the masthead to coincide with the British Army's evacuation.[42] He also kept in contact with printers around the United States—including Patriots such as Isaiah Thomas. With the war ending, for instance, he wrote to Thomas offering to sell him sets of type for his shop (it is unknown whether Thomas accepted the offer).[43] On the other hand, Rivington was forced to drop his newspaper by the end of 1783 once he lost the protection of the British Army.

The willing if not eager presence of the most visible Loyalist in New York after having served the royal government during the war raised questions in many minds that his motivation extended beyond business. One writer speculated that "the sudden transition of Mr. *Rivington* from *his most excellent Majesty's printer*, to being a *republican printer*, and several other circumstances" led him to conclude that Rivington "is still a *printer* to the *British court*, and a *secret emissary*, that is, to watch every opportunity to serve them, at our expence."[44] As specific evidence, the author pointed to Rivington's sale of tickets for an English lottery, which violated republican principles (on the theory that lotteries were harmful to purchasers "unless of some important public service") and transferred money from the United States to Great Britain. Even after he ended his newspaper's run at the end of 1783, other printers still found his continuing presence obnoxious, noting that it "seems to give offence to the present honest republican inhabitants of that metropolis [i.e., New York], and by the last accounts they appear to be determined on his removal."[45] Yet there he remained for a decade more.

For younger printers—even those old enough to have been involved in the political battles of the Revolution—the end of the war brought new opportunities to break out as successful publishers. By the mid-1780s, Isaiah Thomas was doing brisk business as a printer, publisher, and bookseller in Worcester. He owned not only a printing office but also a "Mansion house," considerable land in Worcester, and property in Vermont.[46] By the close of the decade, he had opened an additional bookselling operation in Boston with a former apprentice, Ebenezer Turell Andrews.[47] Nonetheless, Thomas still faced economic hardship. For instance, he had several disputes with other printers and with the Massachusetts state government. In December 1785 he offered sets of type to the Hartford printers Hudson & Goodwin, including "a beautiful font of Minion, &c. neatly dressed and made for my use by Mr. Caslon." Though just arrived and "not yet opened," Thomas wrote that his "demands for Cash are so great" that he felt compelled to sell them in order "to obtain a little of that necessary article."[48] The cash, one might fairly assume, went straight to the importation of still more type—nine thousand dollars worth—not only from Caslon but from other major type founders in London.[49] At the same time, Thomas continued to draw on patronage networks, particularly through John Hancock, who volunteered to pay his bills (via a surrogate) when he admitted to Thomas that "to this moment [I] have pass'd over your Merits unnotic'd."[50] Success in the new republic thus continued to correlate with useful network connections.

Older colonial networks were waning but did not disappear entirely. Even the *eminence gris* of American printing, Benjamin Franklin, found room for one more partnership in the wake of the Revolutionary War. Through this last financial arrangement with a printer during his lifetime, Franklin set up a young printer named Francis Childs. Born in 1763, Childs served his apprenticeship through the years of the Revolutionary War with Philadelphian John Dunlap (who himself had ties to the Franklin network) through the intercession of New York lawyer and politician John Jay. Coming out of his apprenticeship as the war was ending, Childs sought Jay's help to start a printing office in New York City after the British evacuation. He did so both because of Jay's earlier past patronage and because of where Jay was at the time—working at the side of Benjamin Franklin to negotiate peace in Paris. Childs eventually did open an office in New York with types that Franklin acquired for him and continued there for a decade thereafter.[51]

Immigrants, by contrast, often had to build their networks from nothing. Major Atlantic ports received an influx of migrants, among whom were numerous printers, largely from England, Ireland, and Scotland.[52] Mathew Carey easily proved himself the most successful of this group over the course of his career, but the story of his arrival illustrates several common threads.[53] Carey was born in Dublin in 1760 at a time when the Penal Codes restricted the entry of Catholics into many areas of public life. Trained as a printer against the wishes of his father, Carey quickly gained experience courting political and legal controversy during his time in Dublin.[54] He became a strong promoter of Catholic political rights, which drew him to publish pamphlets and a newspaper that excoriated British officials.[55] It also led him to spend a year in exile in France, where he taught school and worked briefly at a press on the recommendation of Benjamin Franklin. Shortly after his return, the Irish Parliament (controlled by Anglicans) had him arrested and held at Newgate Prison in the spring of 1784, but it released him when the session adjourned without adjudicating his case.[56] His detention made clear that his long-term prospects in Dublin were grim, so he departed almost immediately for Philadelphia (dressed, according to legend, in women's garb), a city he chose because he read accounts of his arrest and imprisonment in the *Pennsylvania Packet and Weekly Advertiser*.[57] Lacking formal connections, Carey sought to capitalize on the notoriety he had gained through the transatlantic circulation of news.

With barely a penny to his name, Carey fell into opportunity while still in the midst of the Atlantic crossing. A fellow passenger on the trip introduced him to the Marquis de Lafayette, the young French aide to Washington and hero of the Revolution. Moved by his story and ambitions, Lafayette offered Carey four hundred dollars with which to start his business. Ambitious and brash—traits that would characterize much of his career—he immediately established an office and set out to start a newspaper, the *Pennsylvania Evening Herald*. Though in retrospect he judged himself too rash, at the time Carey forged ahead without quality printing equipment to start the newspaper in January 1785.[58] With it up and running on a weekly basis, Carey immediately sought to leverage that regular presence in public debate to expand his connections and earn new business. He started with two elements of his identity as an Irishman and a printer. Within two months, he had taken on two Irish partners for the newspaper, Christopher Talbot and William Spotswood.[59] He also began to make connections within the printing trade. He undertook the traditional courtesies of the trade, publishing advertisements for other newspaper printers, including Eleazer Oswald of the *New-York Journal* and Francis Childs of the *Daily Advertiser* (New York). He became an agent, taking in subscriptions for the *Maryland Gazette*, published in Baltimore by John Hayes.[60] Finally, he worked to leverage his connection with Lafayette into an endorsement from the most famous and respected American, General George Washington.[61] In so doing, Carey worked to establish a more traditional commercial network of printers with whom he could exchange news, information, and publications.

The Role of the Press in the New Nation

Economic difficulties represented just one area in which printers and their fellow Americans struggled in the aftermath of the Revolutionary War. The printing trade also came to be a focal point as Americans debated how to manage government and society in the new United States. By the 1770s it had become clear that the press served the function of opposition to the imperial government, providing a forum to check its power and authority in the colonies and as a safeguard against tyranny. With the removal of British governance and its replacement with fourteen state governments (including Vermont) and a Congress as an umbrella organization, that function no longer seemed clear for the press. The success of printers at positioning themselves as a key part of society in the colonies during the imperial crisis, in other words, left open the question of how they would best serve society in the new United States.

In their rhetoric, many political leaders continued to see press freedom as a crucial right, though there was rarely debate about what precisely that might mean. Eight states, for example, included freedom of the press among their declarations or bills of rights as they drew up constitutions in the 1770s and 1780s. The North Carolina Constitution contained typical language when it described the press as "one of the great bulwarks of liberty" that "ought never to be restrained."[62] Such provisions reaffirmed the principle that the state would not interfere directly in the productions of the press. Many Americans thus saw the press not only as a necessary safeguard against tyranny—as it had been during the imperial crisis—but as a key component of support for the new republics they had founded.[63] That principle was violated in many ways over the ensuing decades, and debates about the proper role of the press continued well into the early Republic (not to mention the present).[64]

Lawmakers asserted their belief in press freedom but did not hesitate to use the trade as a source of revenue. Like their forebears during the imperial crisis, printers saw additional taxation on printed matter not simply as a commercial burden but more broadly as a threat to the freedom of the press. In 1785 that threat came from an unlikely place: Massachusetts. Struggling to meet its financial obligations, the General Court that year passed a stamp tax on newspapers—the very same sort of tax that had ignited continent-wide protests some twenty years prior, approved unironically by some of the very same men who had disdained the British legislation. Printers reacted as anyone who remembered 1765 could have anticipated: they launched protests across the state and promoted their plight around the nation using their commercial networks.[65] In opposing both the stamp duty and the advertising tax, printers recycled the political strategies and rhetorical arguments employed twenty years earlier. Only a few printers in 1785 had been in business in 1765, but their number included Benjamin Edes, once a founder of the Loyal Nine and the Sons of Liberty and still the publisher of the *Boston Gazette*, and his former partner John Gill, who by the mid-1780s was publishing the *Continental Journal* on his own in Boston. Isaiah Thomas, now well established in Worcester as publisher of the *Massachusetts Spy*, had taken an active role against the British Stamp Act in the less-than-hospitable environment in Halifax. Now these men, along with their younger counterparts new to the intricacies of protesting direct threats to their businesses, turned their attention to an unexpected adversary—the Massachusetts General Court.

In their protests, printers portrayed their newspapers as public services that provided news and "political intelligence" to a broad swath of the population. Their ability to do so, they argued, relied on charges for advertisements, which composed a significant proportion of their revenue (as much as two-thirds).[66] In a petition to the General Court, John Russell noted that advertising revenue "enabled him to afford his papers at such a moderate price, that it easily circulated among all ranks of people, even among those of the lowest fortune." The tax, he believed, would reduce revenue from advertisements and force an increase in subscription costs, which would "prevent the circulation of that political Intelligence, which is manifestly necessary to the virtue, freedom and happiness of the people."[67] Subscribers were notoriously lax about paying, and printers often felt helpless to control that side of their business, but they had more success with advertisers.[68] Furthermore, several printers noted, as John Gill did in the *Continental Journal*, that the act would have unintended consequences—most problematically that it would "encourage our sister states to send their papers in to this Commonwealth cheaper than can possibly be afforded here."[69] Whether their assessment had merit or not, they made the issue a matter of state pride and argued that Massachusetts printers should serve news readers in Massachusetts.

Outside the state, printers without a direct monetary interest portrayed the debate in more starkly ideological terms. In the *Pennsylvania Evening Herald*, Mathew Carey published several essays on freedom of the press. The first, published on August 3, 1785, was widely reprinted around the United States as an expression of freedom of the press and a defense of the Massachusetts printers. In the brief essay, Carey implored "every man in the thirteen states" to "pour out incessant execrations on the devoted heads of those miscreants in Massachusetts" who had passed the act. Carey invoked freedom of the press only in the broadest and most abstract terms, calling it the "palladium of all the rights, privileges, and immunities, dear or sacred to any body of men worthy to rank above the brute creation."[70] In a follow-up essay, Carey portrayed the tax as another assertion of imperial authority against the liberty of the press, arguing that there was no difference among a "Spanish *Licerciado*, a French *Privilege du Roi*, an Oxford *Imprimatur*, an English *Stamp*, and a Massachusetts' *Advertisement Tax*." He reiterated Massachusetts printers' arguments that there was little to be gained from the tax, noting that the stamp tax could bring in only a little more than five hundred pounds per year—not enough, he argued, to pay for the infrastructure required to administer and collect the tax. The tax, therefore, "has been devised solely for the purpose of placing a curb on free discussion." Here he relied on his experience in Dublin where, he argued, the stamp duty placed on newspapers did not pay for the expense of imposing it. This piece, which went into far greater detail at five times the length of the first essay, received reprinting treatment only in Massachusetts and a few neighboring New England states.[71]

Carey's interventions injected a strong Atlantic element into the debate over the Massachusetts advertising tax. Most American printers, including those who worked in Massachusetts in the 1780s, brought an Anglo-American perspective to the issue. For them, the only relevant antecedent was the British Stamp Act of 1765, and by 1785 that was either a distant memory or something that had happened to older colleagues. Carey, on the other hand, had a broader European perspective and more importantly had faced a more pernicious and more direct threat to publishing as a Catholic printer in Anglican-ruled Dublin—strong enough, of course, that the regime had forced him to flee Dublin, not once but twice, and landed him in Philadelphia to comment on affairs in Massachusetts. He had a long record of defending the freedom of the press. That perspective explains why he might compare the tax in Massachusetts to the oppressive print regimes in Spain and France (and, by insinuation, in England). Printers' opposition to the new stamp act ultimately forced the Massachusetts legislature to back down, though it did so slowly and in stages. Within a few months—and without the law ever taking effect—the General Court eliminated the tax on newspapers but left standing that on advertisements. That mollified some, but printers still considered the advertisements tax a significant burden. However, because the legislature did not meet for months at a time, it took until early in 1787 for the advertisements tax to end.

For government officials, the philosophical position that the press should be "free and open" did not mean that printers would be free from prosecution for their publications. Several faced libel charges during the 1780s as Americans established how to balance the principle of a free press with libel laws at a time when criticism of the government was often seen as tantamount to undermining its basis.[72] Eleazer Oswald held the dubious distinction of facing prosecution twice. The son-in-law of John and Elizabeth Holt of New York, Oswald published the *Independent Gazetteer* in Philadelphia and briefly operated the New York printing office after John Holt's death. He was also a bit of a hothead who frequently jousted with others, including Carey, with whom he dueled in 1786 over an escalating series of charges.[73] Another quarrel involved Thomas McKean, the powerful chief justice of Pennsylvania. In 1783 McKean attempted to have Oswald indicted on libel charges for publishing criticism of him, but a grand jury twice declined.[74] Five years later Oswald faced a lawsuit by Andrew Brown, the printer of the *Federal Gazette*, who had demanded that Oswald reveal the names of those who had attacked Brown in Oswald's paper.[75] No one thought that the principle of a free press should be absolute, and Americans spent considerable energy debating where the right ended and seditious libel began.

Printers devoted most of their energies to success at the local level as they struggled to keep their printing offices open. Nonetheless, several turned their eyes to building a broader national culture in which they would serve as the key arbiters of cultural taste through their communications networks. The Continental Congress undertook some of the work to create national symbols and language during and after the Revolutionary War, working in part with printers.[76] Printers, editors, and booksellers began to work out the contours of what would constitute a national literature and communications network. Some argued that politics was the surest way to ensure both nationalist sentiment and financial success. For Noah Webster, the path to a national print culture lay in schoolbooks and language. He saw literature as the "engine of moral progress," which led him to found a newspaper to support the Washington administration as well as a literary magazine during the early republic.[77] And, as the author of the most popular speller in the new United States, Webster had an unparalleled platform to influence the creation of an American language and linguistics.[78] He intended to use it to simplify the language of overwrought English spellings, proposing to eliminate compound vowels such as the "ea" in "bread" or the "u" in words that in British English still have an "-our" ending. In this endeavor (endeavour?) he was only somewhat successful.

As with the efforts of Isaiah Thomas and Robert Aitken in the mid-1770s, several printers tried their hands at a literary magazine as a means to collect and create American culture. Of these, Mathew Carey's *American Museum* (published from 1787–92) remains the most famous. Carey saw the *American Museum* as a distinctly nationalist publication and set about to promote it in precisely that way. He sought subscribers across the nation and worked to build a network of agents to help him accomplish that goal, though over time his efforts fell short.[79] To buttress his subscription numbers, Carey undertook the standard practice of printers promoting a new publication, seeking prominent Americans to endorse the work and, more importantly, to agree to subscribe (or be listed as subscribers if they would permit Carey to send them the *Museum* gratis). As with his *Pennsylvania Evening Herald*, Carey started with George Washington, who happily obliged. The retired general was fulsome

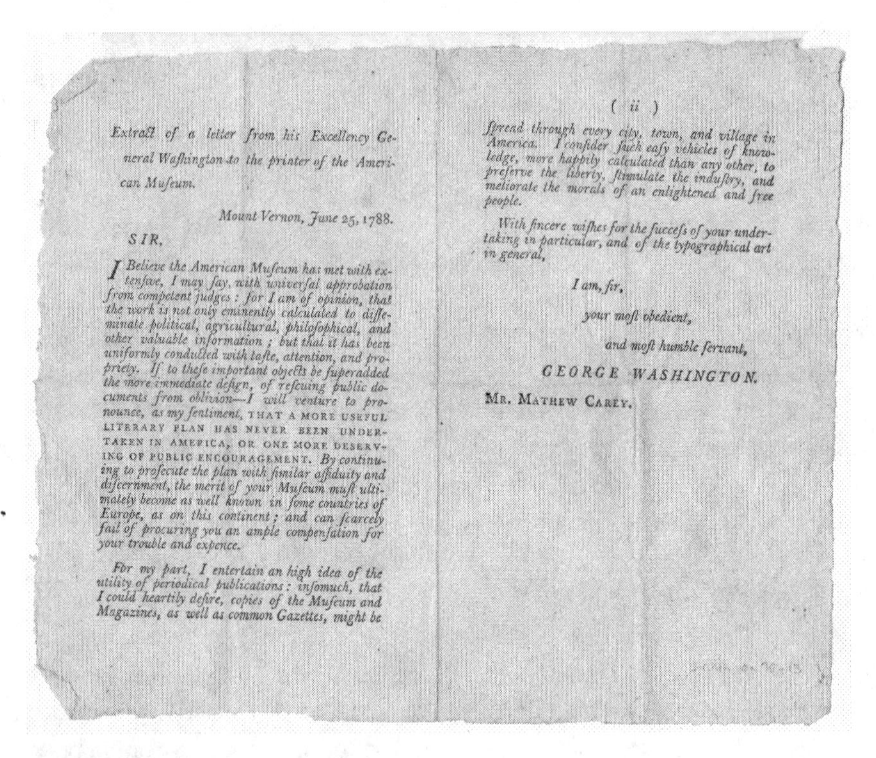

Extract of a letter from his Excellency General Washington to the printer of the American Museum.

Mount Vernon, June 25, 1788.

SIR,

I Believe the American Museum has met with extensive, I may say, with universal approbation from competent judges: for I am of opinion, that the work is not only eminently calculated to disseminate political, agricultural, philosophical, and other valuable information; but that it has been uniformly conducted with taste, attention, and propriety. If to these important objects be superadded the more immediate design, of rescuing public documents from oblivion—I will venture to pronounce, as my sentiment, THAT A MORE USEFUL LITERARY PLAN HAS NEVER BEEN UNDERTAKEN IN AMERICA, OR ONE MORE DESERVING OF PUBLIC ENCOURAGEMENT. *By continuing to prosecute the plan with similar assiduity and discernment, the merit of your Museum must ultimately become as well known in some countries of Europe, as on this continent; and can scarcely fail of procuring you an ample compensation for your trouble and expence.*

For my part, I entertain an high idea of the utility of periodical publications: insomuch, that I could heartily desire, copies of the Museum and Magazines, as well as common Gazettes, might be

(ii)

spread through every city, town, and village in America. I consider such easy vehicles of knowledge, more happily calculated than any other, to preserve the liberty, stimulate the industry, and meliorate the morals of an enlightened and free people.

With sincere wishes for the success of your undertaking in particular, and of the typographical art in general,

I am, sir,

your most obedient,

and most humble servant,

GEORGE WASHINGTON.

MR. MATHEW CAREY.

Figure 13. "Extract of a letter from his Excellency General Washington to the printer of the American Museum" (Philadelphia: Mathew Carey, 1788). Library Company of Philadelphia

in his praise: "A more useful literary plan," he wrote to Carey, "has never been undertaken in america, or one more deserving of public encouragement."[80] Carey found the letter so useful that he printed an excerpt as a broadside to distribute to others—including John Dickinson, from whom he also sought an endorsement (fig. 13).[81] Carey continued to lavish Washington with praise as an avatar of nationalism, dedicating the fourth volume of the magazine (which covered July to December 1788) to him as he was about to be elected president.[82]

In addition to Washington, Carey also received endorsements from other prominent politicians, including Governor William Livingston of New Jersey, who sent Carey material to publish, and Charles Pinckney of South Carolina. With the federal government installed in New York, the July 1789 subscribers' list devoted much of the first page to promoting the patronage of new federal politicians, including President Washington, six senators, and thirteen congressmen.[83] Within the pages of the *Museum*, Carey attempted to encapsulate the new American nationalism he wished to promote. One way he did that was to reprint key texts from the imperial crisis and Revolution to introduce them to new audiences. Beginning in September 1788, for instance, he reprinted the *American Crisis,* Thomas Paine's series of essays first published in late 1776 after the loss of New York City. He also reprinted the letters of the Pennsylvania Farmer, the twelve essays penned by John Dickinson, starting in October 1788. These reprints were part of a larger American project to cement the memory of the Revolution as a moment of national founding, coinciding with, for instance, David Ramsay's history of the American Revolution.[84] The magazine, however much its premise may have captivated the minds of nationalists, failed to achieve the national audience that Carey sought. Though trade networks worked very well for news, they were not yet sophisticated enough in the 1780s to manage large-scale distribution of a national publication.[85]

With the dust settled from the Revolutionary War, the younger generation of American printers began to take bolder stands on issues, to expend less energy on masking their own views, and to espouse more openly a political slant to the news. For the older generation, this was anathema. In a section of his autobiography written in Philadelphia in 1788, Benjamin Franklin lamented the new state of affairs. Printers, he wrote, "make no scruple of gratifying the Malice of Individuals by false Accusations of the fairest Characters among ourselves, augmenting Animosity even to the producing of Duels, and are moreover so indiscrete as to print scurrilous Reflections on the Government of neighbouring States, and even on the Conduct of our best national Allies, which may be attended with the most pernicious Consequences."[86] By not carefully curating the views that made it into the public prints, Franklin argued, printers might ruin the new nation.

Though many young printers had already begun to idolize Franklin as the father of American printing, they paid little heed to his advice. In the first years after the Revolutionary War, printers worked to establish themselves as the key conduits of a burgeoning national communications infrastructure. They moved in large numbers with their fellow Americans into the continental interior, where they established the first printing offices in dozens of new towns. These printers often faced enormous commercial difficulties, as they worked in areas not ready to support printing without significant subsidies. Along the coast, printers faced the inverse problem of intense competition. Younger printers tried to dislodge more experienced artisans, and all battled to drape themselves in the legacy of the Revolution. Throughout the United States, printers reshaped ideas about freedom of the press from a guarantor against imperial tyranny to the necessary accompaniment of liberty in the new republic. As Americans worked to determine the underpinnings of national society and culture, printers positioned themselves to moderate and mediate discussion both in their communities and across the nation.

Conclusion

The publication of the proposed Constitution in September 1787 opened up fundamental debates about the future of the United States. Over the ensuing year, conventions met in each of the thirteen states to determine whether to ratify the Constitution, as per the instructions in Article VII. Just as they had for decades, printers opened the pages of their newspapers to dozens of news articles tracking the state conventions and essays debating the merits of the Constitution. Several writers published series of essays that circulated across the United States, including most famously the *Federalist* (written under the pseudonym Publius by Alexander Hamilton, James Madison, and John Jay), Fabius (John Dickinson), Centinel, Brutus, and Landholder.[1] Those who made arguments in favor of the Constitution made a forceful case for ratification that hinged on the necessity of national unity. The *Federalist* in particular aimed to project a sense of a nation that did not yet exist, using the form of the newspaper to project onto the Constitution a power it would not have for many decades. More than simply a framework for government, Hamilton, Madison, and Jay argued, the Constitution served as a skeleton around which the nation could be built. And by publishing the essays in newspapers and addressing the Constitution logically moving through the seven articles point by point, the corporate author Publius could portray the Constitution as a more coherent document than it actually was, rendering the compromises and half-measures as parts of a logical whole.[2]

Unlike the federalists, who engaged in coordinated campaigns in each state, ratification opponents collaborated almost entirely in print and disagreed about major reasons to prevent the adoption of the Constitution.[3] In general, however, they were concerned that the Constitution would concentrate too much power in the federal government, did not protect the rights of the people, diminished power at the state level, provided insufficient separation of powers, and endowed the executive with unreasonable power.[4] Richard Henry Lee, for example, argued in a pamphlet he signed as the "Federal Farmer" that the federal government envisioned by the Constitution would require that "a multitude of officers and military force be continually kept in view, and employed to enforce the execution of the laws, and to make government feared and respected."[5] Their arguments did not circulate well; by one calculation, only about a dozen newspapers published a significant body of critiques. The first criticism of the Constitution appeared in Philadelphia's *Freeman's Journal* on September 26, 1787, just nine days after the signing. Its critique was mild, according to Pauline Maier, but nonetheless its printer, Francis Bailey, received a "vicious denunciation" in response.[6] For printers opposed to the Constitution, the situation only deteriorated during the ratification debates over the next nine months.

These debates, though they occurred across the nation, did not circulate as much as earlier Revolutionary debates had, such as those over the Stamp Act or Intolerable Acts. Instead, much of the debate about the Constitution remained local. Unlike, say, the letters of the Pennsylvania Farmer in 1767 and 1768, very few of the essay series published about the Constitution circulated that widely, and none approached the level of saturation that Dickinson had achieved as the Farmer. Among the dozen or so essay series, for example, only five individual essays were reprinted in at least twenty newspapers, at a time when nearly one hundred were in print.[7] In large measure this can be attributed to the simple fact that ratification was not one process but thirteen. Each state operated on its own timetable to convene a ratification convention, and though there were common threads to the arguments across states, each had its own mix of issues and debates come to the fore.[8] Arguments and discussion nonetheless circulated around the United States about the Constitution as soon as delegates signed the document on September 17, 1787. Newspapers tracked the progress of ratification as states began to vote in December and into the spring of 1788. And they generally showed interest in the progress of the debates, if not the substance of the legal and philosophical arguments.

Scholars have usually portrayed printers as almost uniformly in favor of the Constitution. For various reasons, especially the strong correlation between printing and cities, most printers were indeed federalists—that is, pro-ratification. The trade thrived in larger towns and cities where a critical mass of readers could purchase and consume printed material. By 1787, most major American cities relied on commerce and trade—endeavors that would benefit from the new Constitution—and so naturally printers sided with the interests of their localities. Before and during the Constitutional Convention, for instance, many printers published essays and news items that underscored what they believed to be the necessity of the convention and reforms to the structure of the national government. Once the ratification process began, pro-Constitution arguments dominated newspaper coverage, far out of proportion to the support for the Constitution among the American population.[9]

The rhetoric of printers as impartial actors persisted nonetheless. Nearly every printer in the United States continued to assert his or her commitment to a "free and open press," the same phrase that they and their forebears had used for decades. In the *New-York Journal*, for example, Eleazer Oswald, the son-in-law of Revolutionary printer John Holt and inheritor of his newspaper, wrote that he "professes to print an *impartial* paper" and that he would "ever act AS A PRINTER, giving to every performance, that may be written with decency, free access to his Journal."[10] In a similar fashion, the editor of the *United States Chronicle*, published in Providence, printed a letter that argued that the "Liberty of the Press" was "a Privilege of infinite Importance." The writer even invoked the American Revolution, suggesting that Americans had "fought and bled" for that right and the he "would again shoulder my Musket" to defend it.[11] Fissures in the ideology of the free and open press, however, were beginning to develop. Slowly but surely during the ratification debates, printers let slip both that they possessed opinions and ideas of their own and that they saw themselves as active decision makers as to what content should appear in their newspapers. In the same paragraph as he proclaimed his impartiality, for instance, Oswald noted that he acted in his professional capacity while "setting aside his private political sentiments."[12] By implication, therefore, another printer—or even Oswald at another time—might not set aside their sentiments.

Following his conception of American patriotism, one Boston printer firmly declared that the tradition of anonymity no longer applied. Benjamin Russell, publisher of the *Centinel*, announced in October 1787 that he would not run any essays that opposed the Constitution unless the author was willing to reveal his or her name to the public to allay general fears that foreign agents or unpatriotic elements were behind such arguments. Other nations, one writer suggested in the *Boston Independent Chronicle*, wanted to "to keep us in our divided and distracted condition" and consequently would "probably fill the press with objections against the report of the Convention."[13]

Another writer in the *Philadelphia Independent Gazetteer* suggested that the printer should publish names so that he could "declare that every writer is either a NATIVE or a CITIZEN of one of these states."[14] They also implied sinister motives to anyone who hesitated to attach his or her name to a public argument about the Constitution. Editors described those who opposed ratification as an "antifederal junto" whose "cause is that of the devil" and wrote that those who wrote against the Constitution were "enemies and traitors to their country."[15]

Those newspapers whose editors had antifederalist inclinations, meanwhile, argued that most of their peers were suppressing arguments opposed to ratification. In the view of these printers, attempts to reveal the names of authors violated the liberty of the press as part of a broader effort to disparage those who opposed the Constitution. In the *Freeman's Journal*, for example, Francis Bailey alleged that pro-Constitution forces equated "an *anti-federalist* and a *tory*," so that antifederalists stood rhetorically as opponents not merely of the Constitution but of the Revolution itself.[16] Anti-ratification printers rekindled the revolutionary rhetoric of slavery and freedom to rebut the charge. In a letter addressed to Thomas Greenleaf, editor of the *New-York Journal*, Detector argued that Benjamin Russell's policy of revealing names was akin to slavery. "The printers of a free community," Detector wrote, "are an important set of men—and, when *they* league to enslave it—it will be enslaved indeed."[17] Finally, they made an argument that seems curious for a group of printers, that public opinion—touted by federalists as a reason to reveal the names of authors—was an unreliable barometer. "A man of sense," one writer argued in the *Freeman's Journal*, "expects some other proof of a paper being *impious, heretical*, or *treasonable*, than merely that of its being burned by the hands of the common hangman."[18] In addition to a rhetorical barrage, printers who published against the Constitution also faced violence at the hands of pro-ratification forces. Thomas Greenleaf, for example, the publisher of the *New-York Journal* saw his office destroyed on July 26, 1788, the very night that the New York convention voted in favor of ratification.[19]

The adoption of the Constitution failed to quell the debate over newspaper circulation and the proper role of the post office in national communications. The intensity of the ratification debate, in fact, led some printers to question the effectiveness and impartiality of the newspaper exchange policy. As the states were debating ratification in the winter of 1788, some antifederalists charged that the post office, controlled by the Federalist postmaster general, Ebenezer Hazard, was preventing their newspapers from being transported among the states and thus depriving delegates to the state conventions of access to their arguments against the Constitution. Eleazer Oswald pushed the issue when he published Centinel's essay number XI in which the author suggested that postmasters and post riders, "in violation of their duty and integrity have prostituted their of—ces to forward the nefarious design of enslaving their countrymen, by thus cutting off all communication by the usual vehicle between the patriots of America."[20] For antifederalists, having the post office at their disposal was crucial in order for them to circulate their dissenting arguments.[21]

Even William Goddard, the old adversary of the despotic British postal system, entered the fray. In February 1788 he wrote to Philadelphia publisher Mathew Carey to suggest that once again a new post office might be founded, this time in opposition to the one run by Congress. He also argued publicly that the postal system's policies restricted the free flow of newspapers through the states and reminded readers of the *Maryland Journal* that "a similar Measure, previous to the American Revolution, was very severely reprobated and resented throughout the Continent, as having a manifest Tendency to endanger Public Liberty, (as well as greatly to injure Individuals) by shutting up the Channels of Public Information." He closed with a warning: "The present Post-Office Administration would do well to reflect on the Fate of their Predecessors."[22] The statement was indeed incendiary, especially for a man whose sister served as postmistress in Baltimore, their shared city of residence. Printers in Philadelphia rallied in favor of the free circulation of exchange papers, petitioning the Pennsylvania Assembly in March 1788 to encourage its delegates to Congress to fight for the "the restoration of this their necessary and long accustomed privilege."[23]

Because printers as a group largely supported the new Constitution, opposition to the post office fizzled outside of the small coterie of antifederalists who shared complaints about the distribution of patronage among their enemies.[24] Unlike the 1770s, when the Boston Committee of Correspondence could generate public disgust over the imperial post office, castigation of the Continental Post Office as an evil entity now had little rhetorical appeal. The Continental Congress, meanwhile, clarified and doubled down on its policies. In May 1788 Congress passed a resolution that reaffirmed its previous position that printers were "allowed to exchange their papers with each other by means of the public mail without any charge of postage," so long as the newspaper was not being used to conceal any other letters or newspapers.[25] These debates finally died out in 1792 when the second Federal Congress passed a Post Office Act that provided a broad and uniform policy for the shipment of newspapers through the mail. In so doing, it set the stage for the rapid expansion of the American communications infrastructure.[26] By establishing a standard policy, the 1792 act ended disputes over the post office that had run through the entire eighteenth century and effectively established newspapers as the main source of news in the young nation.

The Constitution formally achieved ratification when New Hampshire became the ninth state to vote in favor of the plan of government on June 21, 1788 (followed a few days later by Virginia, whose delegates thought they had won the right to put the document in force until news arrived from New Hampshire).[27] Printers claimed a place of prominence in the celebrations of the Constitution's ratification and, in the process, offered a glimpse of an inchoate guild mentality. Most famously, printers participated actively in the Grand Processions that many cities and towns held to commemorate ratification. In Charleston, for example, the printers of the town marched as a trade "with a press, frames and cases on a stage drawn by horses.—Compositors and pressmen at work."[28] In the newspaper's account, most tradesmen simply marched with their compatriots, but printers showed themselves practicing their trade. Printers thus portrayed themselves, as they had for several years, as central to the effort of the Revolution. Raised in an era of political dispute, the printers who came of age during and after the Revolution were far less bashful about their beliefs and far more willing to use their presses as the engines of party politics.

Printers also kept the issue of press freedom salient in the national conversation as first the state ratification conventions and then the new Federal Congress considered whether to amend the Constitution with a bill of rights. In a letter published in January 1789, for example, James Madison (then a candidate for the House of Representatives from Virginia) argued that he prioritized ratification above all else. With that question settled, he sought to offer amendments, "guard essential rights," and "render certain vexatious abuses of power," including protections for freedom of the press.[29] Once in Congress, Madison drafted amendments to the Constitution that in their original form would have added language to Article I to protect the freedom of the press from intrusion by both federal and state governments.[30] Congress ultimately chose to append amendments rather than to change the language of the original seven articles and included freedom of the press among several others intended to guarantee freedom of conscience and expression.[31]

In the 1790s, this tendency would help printers and newspaper editors to become central players in the early national political parties—a story covered well by historians in the past fifteen years. Along with major political figures such as James Madison, many printers swung from support for the Constitution and a more centralized government in the late 1780s to suspicions about the practices of the new federal government in the 1790s. A core of these printers thus came to form the backbone of the Republican party of Jefferson and Madison. These national politicians in turn fostered and supported a lively oppositional press in the 1790s, in part by hiring and retaining printers and editors in government positions as ways to prop them up so they could circulate favorable news and essays. The Post Office Act of 1792 encouraged the establishment of newspapers and the post office as the key components of the political communications infrastructure for the new nation by setting cheap rates of postage for newspapers (and, by contrast, high rates for personal letters).[32]

The design of the infrastructure thus fostered the development of the press as a partisan engine, especially in larger cities that could support numerous papers. (These developments happened concurrently in England.)[33] No target was above reproach, not even George Washington, the most popular and most admired man in the new United States. By the time he left office in March 1797, Republican editors had mercilessly hacked at him, his policies, and his administration for years. No longer did the principle of an outwardly nonpartisan press hold. At the vanguard of this partisan assault was none other than Benjamin Franklin Bache—the eponymous grandson of the man who created the "free and open press" doctrine. On the day that John Adams was inaugurated as the second president of the United States, Bache (no friend of Adams) published an essay in his *Aurora* newspaper that the day was a "JUBILEE" for the United States because "the name of WASHINGTON from this day ceases to give a currency to political iniquity; and to legalize corruption."[34] No longer standing outside the fray as "meer mechanics," printers set themselves instead as the key combatants in the politics of the early United States.

Abbreviations

AAS	American Antiquarian Society, Worcester, MA
APS	American Philosophical Society, Philadelphia
BCC Papers	Boston Committee of Correspondence Records. Manuscripts and Archives Division, New York Public Library, Astor, Lenox and Tilden Foundations, New York
CBAW	Hugh Amory and David D. Hall, eds. *The Colonial Book in the Atlantic World.* Vol. 1 of *A History of the Book in America.* Cambridge: Cambridge University Press, 2000.
DHRC	John P. Kaminski et al., eds. *The Documentary History of the Ratification of the Constitution Digital Edition.* Charlottesville: University of Virginia Press, 2009. http://rotunda.upress.virginia.edu/founders/RNCN.html, vols. 13–18
DLAR	David Library of the American Revolution, Washington Crossing, PA
HSP	Historical Society of Pennsylvania, Philadelphia
JCC	Worthington Chauncey Ford, ed. *Journals of the Continental Congress*, 34 vols. Washington, DC: Government Printing Office, 1904–37.
LCP	Library Company of Philadelphia, Philadelphia
NYPL	Manuscripts and Archives Division, New York Public Library, Astor, Lenox, and Tilden Foundations, New York
N-YHS	New-York Historical Society, New York
PBF	Leonard W. Labaree et al., eds. *The Papers of Benjamin Franklin*, 43 vols. New Haven, CT: Yale University Press, 1959 to the present.

Introduction

1. Isaiah Thomas, *Three Autobiographical Fragments; Now First Published upon the 150th Anniversary of the Founding of the American Antiquarian Society, October 24, 1812* (Worcester, MA: American Antiquarian Society, 1962), 14.

2. William Goddard to Isaiah Thomas, undated, April 15 and April 22, 1811; William McCulloch to Isaiah Thomas, September 1, 1812; Isaiah Thomas Papers, AAS. See also Clarence S. Brigham, "William McCulloch's Additions to Thomas's History of Printing," *Proceedings of the American Antiquarian Society* 31, no. 1 (April 1921): 89–100.

3. Isaiah Thomas, *The History of Printing in America, with a Biography of Printers, and an Account of Newspapers*, 2 vols. (Worcester, MA: Isaiah Thomas, Jr., 1810), 1:9.

4. Thomas, *History of Printing in America*, 1:15.

5. Thomas, *History of Printing in America*, 1:342.

6. Thomas, *History of Printing in America*, 2:314.

7. Thomas, *History of Printing in America*, 2:315.

8. David Ramsay, *The History of the American Revolution* (Philadelphia: Printed and sold by R. Aitken & Son, 1789), Early American Imprints, 1st ser., no. 22090, 2:319.

9. Thomas, *History of Printing in America*, 1:209.

10. Marcus L. Daniel, *Scandal and Civility: Journalism and the Birth of American Democracy* (New York: Oxford University Press, 2009); Trish Loughran, *The Republic in Print: Print Culture*

in the Age of U.S. Nation Building, 1770–1870 (New York: Columbia University Press, 2007); Jeffrey L. Pasley, *"The Tyranny of Printers": Newspaper Politics in the Early American Republic*, Jeffersonian America (Charlottesville: University Press of Virginia, 2001); Saul Cornell, *The Other Founders: Anti-Federalism and the Dissenting Tradition in America, 1788–1828* (Chapel Hill: OIEAHC, University of North Carolina Press, 1999); David Waldstreicher, *In the Midst of Perpetual Fetes: The Making of American Nationalism, 1776–1820* (Chapel Hill: OIEAHC, University of North Carolina Press, 1997); Rosalind Remer, *Printers and Men of Capital: Philadelphia Book Publishers in the New Republic* (Philadelphia: University of Pennsylvania Press, 1996). For the British side during the Revolution, see Troy Bickham, *Making Headlines: The American Revolution as Seen through the British Press* (DeKalb: Northern Illinois University Press, 2009); Eliga H. Gould, *The Persistence of Empire: British Political Culture in the Age of the American Revolution* (Chapel Hill: OIEAHC, University of North Carolina Press, 2000); Kathleen Wilson, *The Sense of the People: Politics, Culture, and Imperialism in England, 1715–1785* (Cambridge: Cambridge University Press, 1995). One exception which provides a frustratingly presentist account of revolutionary-era newspapers is William B. Warner, *Protocols of Liberty: Communication Innovation and the American Revolution* (Chicago: University of Chicago Press, 2013).

11. On the importance of access to information in colonial and Revolutionary America, see Richard D. Brown, *Knowledge Is Power: The Diffusion of Information in Early America, 1700–1865* (New York: Oxford University Press, 1989). See also William Slauter, "News and Diplomacy in the Age of the American Revolution" (PhD diss., Princeton University, 2007); Ian K. Steele, *The English Atlantic, 1675–1740: An Exploration of Communication and Community* (New York: Oxford University Press, 1986).

12. Hugh Amory, "A Note on Statistics," in *CBAW*, 512.

13. The size of paper varied widely, so the numbers are a rough approximation and generalization.

14. The description of printing a newspaper in this paragraph owes a great deal to Vincent Golden, curator of newspapers and periodicals at AAS. The process of printing a single page involved thirteen distinct processes. Carol Sue Humphrey, *"This Popular Engine": New England Newspapers during the American Revolution, 1775–1789* (Newark: University of Delaware Press, 1992), 27.

15. On advertising and print during the imperial crisis, see T. H. Breen, *The Marketplace of Revolution: How Consumer Politics Shaped American Independence* (New York: Oxford University Press, 2004); Carl Robert Keyes, "Early American Advertising: Marketing and Consumer Culture in Eighteenth-Century Philadelphia" (PhD diss., Johns Hopkins University, 2007).

16. See *Georgia Gazette*, April 7, 1763; *South-Carolina Gazette*, April 30, 1763; *Boston Evening-Post*, May 16, 1763; *Newport Mercury*, May 16, 1763; *New-York Gazette or Weekly Post-Boy*, May 19, 1763; *New-York Mercury*, May 23, 1763.

17. Uriel Heyd, *Reading Newspapers: Press and Public in Eighteenth-Century Britain and America*, SVEC 2012:3 (Oxford: Voltaire Foundation, 2012).

18. *Quebec Gazette*, June 21, 1764.

19. *Rivington's New-York Gazetteer*, April 22, 1773.

20. Victoria E. M. Gardner, *The Business of News in England, 1760–1820*, Palgrave Studies in the History of the Media (Houndmills: Palgrave Macmillan, 2016); Hannah Barker, *Newspapers, Politics, and Public Opinion in Eighteenth-Century England* (Oxford: Clarendon Press, 1998); Jeremy Black, *The English Press in the Eighteenth Century* (London: Croom Helm, 1987); Michael Harris, *London Newspapers in the Age of Walpole: A Study of the Origins of the Modern English Press* (Rutherford, NJ: Fairleigh Dickinson University Press, 1987); Wilson, *The Sense of the People.*

21. Victor Hugo Paltsits, "New Light on 'Publick Occurrences': America's First Newspaper," *Proceedings of the American Antiquarian Society* 59, no. 1 (1949): 75–88; Charles E. Clark, "Early American Journalism," in *CBAW*, 351; Steele, *English Atlantic.*

22. Charles E. Clark, *The Public Prints: The Newspaper in Anglo-American Culture, 1665–*

1740 (Oxford: Oxford University Press, 1994); Wm. David Sloan and Julie Hedgepeth Williams, *The Early American Press, 1690–1783*, vol. 1 of *The History of American Journalism* (Westport, CT: Greenwood Press, 1994); David A. Copeland, *Colonial American Newspapers: Character and Content* (Newark: University of Delaware Press, 1997).

23. Alejandra Dubcovsky, *Informed Power: Communication in the Early American South* (Cambridge, MA: Harvard University Press, 2016); Katherine Grandjean, *American Passage: The Communications Frontier in Early New England* (Cambridge, MA: Harvard University Press, 2015); Lindsay O'Neill, *The Opened Letter: Networking in the Early Modern British World*, Early Modern Americas (Philadelphia: University of Pennsylvania Press, 2015); Kenneth J. Banks, *Chasing Empire across the Sea: Communications and the State in the French Atlantic, 1713–1763* (Montreal: McGill-Queen's University Press, 2002); Konstantin Dierks, *In My Power: Letter Writing and Communications in Early America* (Philadelphia: University of Pennsylvania Press, 2009).

24. Examples include "Mapping the Republic of Letters," http://republicofletters.stanford.edu/; "Six Degrees of Francis Bacon," http://www.sixdegreesoffrancisbacon.com/; "The Atlas of the Rhode Island Book Trade in the Eighteenth Century," Rhode Island Historical Society, http://www.rihs.org/atlas/index.php.

25. Bernard Bailyn, *The Ideological Origins of the American Revolution*, rev. ed. (Cambridge, MA: Belknap Press of Harvard University Press, 1992); Gordon S. Wood, *The Creation of the American Republic, 1776–1787* (Chapel Hill: OIEAHC, University of North Carolina Press, 1969; repr. 1998); J. G. A. Pocock, *The Machiavellian Moment: Florentine Political Thought and the Atlantic Republican Tradition*, 2nd ed. (Princeton, NJ: Princeton University Press, 2003); Pauline Maier, *American Scripture: Making the Declaration of Independence* (New York: Alfred A. Knopf, 1997); Jay Fliegelman, *Prodigals and Pilgrims: The American Revolution against Patriarchal Authority, 1750–1800* (Cambridge: Cambridge University Press, 1982); Arthur M. Schlesinger Sr., *Prelude to Independence: The Newspaper War on Britain, 1764–1776* (New York: Alfred A. Knopf, 1957); Philip G. Davidson, *Propaganda and the American Revolution, 1763–1783* (Chapel Hill: University of North Carolina Press, 1941); Carl Berger, *Broadsides and Bayonets: The Propaganda War of the American Revolution* (Philadelphia: University of Pennsylvania Press, 1961); Pauline Maier, *From Resistance to Revolution: Colonial Radicals and the Development of American Opposition to Britain* (New York: Alfred A. Knopf, 1972).

26. Breen, *The Marketplace of Revolution*; Brendan McConville, *The King's Three Faces: The Rise & Fall of Royal America, 1688–1776* (Chapel Hill: OIEAHC, University of North Carolina Press, 2006).

27. Eric Slauter, "Reading and Radicalization: Print, Politics, and the American Revolution," *Early American Studies* 8, no. 1 (2010): 5–40. See also Larzer Ziff, *Writing in the New Nation: Prose, Print, and Politics in the Early United States* (New Haven, CT: Yale University Press, 1991); David S. Shields, *Civil Tongues & Polite Letters in British America* (Chapel Hill: OIEAHC, University of North Carolina Press, 1997); Joan Shelley Rubin, "What Is the History of the History of Books?," *Journal of American History* 90 (2003): 555–75; David D. Hall, *Worlds of Wonder, Days of Judgment: Popular Religious Belief in Early New England* (Cambridge, MA: Harvard University Press, 1990); Roger Chartier, *The Cultural Uses of Print in Early Modern France* (Princeton, NJ: Princeton University Press, 1987); Bernard Bailyn and John B. Hench, eds., *The Press and the American Revolution*, (Worcester, MA: American Antiquarian Society, 1980).

28. The five-volume *A History of the Book in America* series, most notably, has opened many new avenues of research to pursue in linking the material processes of printing and the political culture of the British North American colonies and United States. Hugh Amory and David D. Hall, eds., *The Colonial Book in the Atlantic World*, vol. 1 of *A History of the Book in America* (Cambridge: Cambridge University Press, 2000); Robert A. Gross and Mary Kelley, eds., *An Extensive Republic: Print, Culture, and Society in the New Nation, 1790–1840*, vol. 2 of *A History of the Book in America* (Chapel Hill: University of North Carolina Press, 2010).

29. Jürgen Habermas, *The Structural Transformation of the Public Sphere: An Inquiry into a Category of Bourgeois Society*, trans. Thomas Burger (Cambridge, MA: MIT Press, 1989). Since the translation of Habermas's book in 1989, the literature has grown exponentially. Some of the most relevant work includes Michael Warner, *Letters of the Republic: Publication and the Public Sphere in Eighteenth-Century America* (Cambridge, MA: Harvard University Press, 1990); Craig A. Calhoun, ed., *Habermas and the Public Sphere* (Cambridge, MA: MIT Press, 1992); John Brewer, "This, That and the Other: Public, Social and Private in the Seventeenth and Eighteenth Centuries," in *Shifting the Boundaries: Transformation of the Languages of Public and Private in the Eighteenth Century*, ed. Dario Castiglione and Lesley Sharpe (Exeter: Exeter University Press, 1995), 1–21; Richard R. John, *Spreading the News: The American Postal System from Franklin to Morse* (Cambridge, MA: Harvard University Press, 1995); Jeff Weintraub and Krishan Kumar, *Public and Private in Thought and Practice: Perspectives on a Grand Dichotomy* (Chicago: University of Chicago Press, 1997); Christopher Grasso, *A Speaking Aristocracy: Transforming Public Discourse in Eighteenth-Century Connecticut* (Chapel Hill: OIEAHC, University of North Carolina Press, 1999); "Forum: Alternate Histories of the Public Sphere," special issue, *William and Mary Quarterly*, 3rd ser., 62, no. 1 (2005). For a useful counterargument, see Robert A. Gross, "Print and the Public Sphere in Early America," in *The State of U.S. History*, ed. Melvyn Stokes (Oxford: Berg, 2002), 245–64.

30. Warner, *The Letters of the Republic.*

31. Benedict Anderson, *Imagined Communities: Reflections on the Origin and Spread of Nationalism*, rev. ed. (New York: Verso, 1991).

32. Anderson, *Imagined Communities*, 61.

33. Ed White, "Early American Nations as Imagined Communities," *American Quarterly* 56, no. 1 (2004): 49–81.

34. Trish Loughran, *The Republic in Print: Print Culture in the Age of U.S. Nation Building, 1770–1870* (New York: Columbia University Press, 2007).

35. Robert G. Parkinson, *The Common Cause: Creating Race and Nation in the American Revolution* (Chapel Hill: OIEAHC, University of North Carolina Press, 2016).

36. Benjamin E. Park, *American Nationalisms: Imagining Union in the Age of Revolutions, 1783–1833* (Cambridge: Cambridge University Press, 2018); Carroll Smith-Rosenberg, *This Violent Empire: The Birth of an American National Identity* (Chapel Hill: OIEAHC, University of North Carolina Press, 2010); Jonathan Senchyne, "Paper Nationalism: Material Textuality and Communal Affiliation in Early America," *Book History* 19 (2016): 66–85; Robb K. Haberman, "Provincial Nationalism: Civic Rivalry in Postrevolutionary American Magazines," *Early American Studies, An Interdisciplinary Journal* 10, no. 1 (Winter 2012): 162–93; Tim Cassedy, "'A Dictionary Which We Do Not Want': Defining America against Noah Webster, 1783–1810," *William and Mary Quarterly*, 3rd ser., 71, no. 2 (April 2014): 229–54; Benjamin H. Irvin, *Clothed in Robes of Sovereignty: The Continental Congress and the People Out of Doors* (New York: Oxford University Press, 2011).

37. Isaiah Thomas indenture, Isaiah Thomas Papers, AAS.

38. Richard R. John and Jonathan Silberstein-Loeb, eds., *Making News: The Political Economy of Journalism in Britain and America from the Glorious Revolution to the Internet* (Oxford: Oxford University Press, 2015).

Chapter 1 · The Business and Economic World of the Late Colonial Printing Trade

1. "Samuel Hall," *Atlas of the Rhode Island Book Trade in the Eighteenth Century*, Rhode Island Historical Society, http://www.rihs.org/atlas/details.php?id=22.

2. Nathaniel Ames, *An astronomical diary: or, almanack for the year of our Lord Christ,*

1764. . . . Calculated for the meridian of Boston, New-England, lat. 42 deg. 25 min. north (Newport: reprinted by Samuel Hall, 1763), Early American Imprints, ser. 1, no. 9323.

3. The October 24, 1763, edition of the *Newport Mercury* reprinted several paragraphs and letter extracts from the *Pennsylvania Gazette* of October 13, 1763.

4. Stephen Botein, "'Meer Mechanics' and an Open Press: The Business and Political Strategies of Colonial American Printers," *Perspectives in American History* 9 (1975): 127–225.

5. On elite apprenticeship, see Robert Martello, *Midnight Ride, Industrial Dawn: Paul Revere and the Growth of American Enterprise* (Baltimore: Johns Hopkins University Press, 2010).

6. On the identity of "middling" as a cultural construct in early modern England, see Margaret Hunt, *The Middling Sort: Commerce, Gender, and the Family in England, 1680–1780* (Berkeley: University of California Press, 1996). On merchants in the British Atlantic, see Thomas Doerflinger, *A Vigorous Spirit of Enterprise: Merchants and Economic Development in Revolutionary Philadelphia* (Chapel Hill: IEAHC, University of North Carolina Press, 1986); David Hancock, *Citizens of the World: London Merchants and the Integration of the British Atlantic Community, 1735–1785* (Cambridge: Cambridge University Press, 1995); Cathy D. Matson, *Merchants & Empire: Trading in Colonial New York* (Baltimore: Johns Hopkins University Press, 1998); Natasha Glaisyer, *The Culture of Commerce in England, 1660–1720* (Woodbridge, UK: Royal Historical Society / The Boydell Press, 2006). On laboring people in colonial and Revolutionary America, see Billy G. Smith, *The "Lower Sort": Philadelphia's Laboring People, 1750–1800* (Ithaca, NY: Cornell University Press, 1991); Gary B. Nash, *The Urban Crucible: Social Change, Political Consciousness, and the Origins of the American Revolution* (Cambridge, MA: Harvard University Press, 1979); Seth Rockman, *Scraping By: Wage Labor, Slavery, and Survival in Early Baltimore* (Baltimore: Johns Hopkins University Press, 2009). Simon Newman has argued that Franklin in his career maintained a working-class identity even after he achieved fame. "Benjamin Franklin and the Leather-Apron Men: The Politics of Class in Eighteenth-Century Philadelphia," *Journal of American Studies* 43, no. 2 (2009): 161–75.

7. On the importance of family in establishing business connections, see Ellen Hartigan-O'Connor, *The Ties That Buy: Women and Commerce in Revolutionary America* (Philadelphia: University of Pennsylvania Press, 2009), 69–100; Peter Mathias, "Risk, Credit and Kinship in Early Modern Enterprise," in *The Early Modern Atlantic Economy*, ed. John J. McCusker and Kenneth Morgan (Cambridge: Cambridge University Press, 2000), 15–35; Toby L. Ditz, "Formative Ventures: Eighteenth-Century Commercial Letters and the Articulation of Experience," in *Epistolary Selves: Letters and Letter-Writers, 1600–1945*, ed. Rebecca Earle (London: Ashgate, 1999), 59–78; John J. McCusker and Russell R. Menard, *The Economy of British America, 1607–1789* (Chapel Hill: IEAHC, University of North Carolina Press, 1985), 329; Patricia Cleary, "Making Men and Women in the 1770s: Culture, Class, and Commerce in the Anglo-American World," in *A Shared Experience: Men, Women, and the History of Gender*, ed. Laura McCall and Donald Yacovone (New York: New York University Press, 1998), 98–117. On the household as economic unit, see Carole Shammas, *A History of Household Government in America* (Charlottesville: University Press of Virginia, 2002); John Demos, *A Little Commonwealth: Family Life in Plymouth Colony* (New York: Oxford University Press, 1970); Laurel Thatcher Ulrich, *A Midwife's Tale: The Life of Martha Ballard, Based on Her Diary, 1785–1812* (New York: Alfred A. Knopf, 1990); Susan Kellogg and Steven Mintz, *Domestic Revolutions: A Social History of American Family Life* (New York: Free Press, 1988).

8. Joyce E. Chaplin, ed., *Benjamin Franklin's Autobiography*, Norton Critical Edition (New York: W. W. Norton, 2012), 77.

9. Ebenezer Hazard to William Bradford, March 26, 1773, Ferdinand J. Dreer Autograph Collection [coll. 175], HSP.

10. Caitlin DeAngelis Hopkins, "'A Negro Man, by Trade a Silversmith': Enslaved Artisans

in the Northern Anglo-American Colonies, 1700–1785," Omohundro Institute 23rd Annual Conference, Ann Arbor, MI, June 16, 2017.

11. Isaiah Thomas, *The History of Printing in America, with a Biography of Printers, and an Account of Newspapers*, 2 vols. (Worcester, MA: Isaiah Thomas, Jr., 1810), 2:159; David Waldstreicher, *Runaway America: Benjamin Franklin, Slavery, and the American Revolution* (New York: Hill and Wang, 2004).

12. Sidney E. Berger, "Innovation and Diversity among the Green Family of Printers," *Printing History* 12, no. 1 (1990): 2–20; Charles Wetherell, "Brokers of the Word: An Essay in the Social History of the Early American Press, 1639–1783" (PhD diss., University of New Hampshire, 1980), 80.

13. Thomas, *History of Printing in America*, 1:341–43.

14. On marriage as a means of extending business ties, see Richard Grassby, *Kinship and Capitalism: Marriage, Family, and Business in the English-Speaking World, 1580–1740* (Cambridge: Cambridge University Press, 2001); Rosalind J. Beiler, *Immigrant and Entrepreneur: The Atlantic World of Caspar Wistar, 1650–1750* (University Park: Pennsylvania State University Press, 2008), 89–108; Leonore Davidoff and Catherine Hall, *Family Fortunes: Men and Women of the English Middle Class, 1780–1850*, rev. ed. (London: Routledge, 2002); Hancock, *Citizens of the World*, 43; Glaisyer, *The Culture of Commerce*, 13–16.

15. Data cited throughout the book is drawn from a database of 756 master printers active between 1756 and 1796 that I have compiled. For further information, see the Essay on Sources. It is possible that more married within printing families, because not all female spouses could be positively identified as having family members in the printing trade. Of the printers with known marital status, only three never married.

16. Richard L. Demeter, *Printers, Presses, and Composing Sticks: Women Printers of the Colonial Period* (Hicksville, NY: Exposition Press, 1979).

17. On women's administration of estates in the colonial era, see Sara T. Damiano, "'To Well and Truly Administer': Female Administrators and Estate Settlement in Newport, Rhode Island, 1730–1776," *New England Quarterly* 86, no. 1 (2013): 89–124.

18. William Goddard, *The Partnership: Or the History of the Rise and Progress of the Pennsylvania Chronicle, &c. Wherein the Conduct of* JOSEPH GALLOWAY, *Esq; Speaker of the Honourable House of Representatives of the Province of* Pennsylvania, *Mr.* THOMAS WHARTON, *sen. and* their Man BENJAMIN TOWNE, *my late Partners with my own, is properly delineated, and their Calumnies against me fully refuted* (Philadelphia: William Goddard, 1770), Early American Imprints, 1st ser., no. 11669, 20; Susan Henry, "Sarah Goddard, Gentlewoman Printer," *Journalism Quarterly* 57 (1980): 23–30; Nancy Fisher Chudacoff, "Woman in the News, 1762–1770: Sarah Updike Goddard," *Rhode Island History* 32 (1973): 99–105.

19. Christopher J. Young, "Mary K. Goddard: A Classical Republican in a Revolutionary Age," *Maryland Historical Magazine* 96 (2001): 4–27.

20. David D. Hall, "The Atlantic Economy in the Eighteenth Century," in *CBAW*, 155–59.

21. Francis-J. Audet, "William Brown (1737–1789). Premier imprimeur, journaliste et libraire de Québec. Sa vie et ses oeuvres," *Royal Society of Canada Transactions*, 3rd ser., 26, no. 1 (1932): 97–112; André Beaulieu and Jean Hamelin, *La presse québécoise des origines à nos jours*, 2 vols. (Québec: Les presses de l'Université Laval, 1973), 1:1–4; Ægidius Fauteux, *The Introduction of Printing into Canada* (Montreal: Rolland Paper Co., 1930), 71–77.

22. "[Il] décidait d'aller vivre sous un climat plus hospitalier." Audet, "William Brown," 97.

23. The *Gazette* continued publication until 1874. Audet, "William Brown," 111.

24. For a more detailed examination of printers and immigration, see Joseph M. Adelman, "Transatlantic Migration and the Printing Trade in Revolutionary America," *Early American Studies* 11, no. 3 (Fall 2013): 516–44.

25. This number includes all printers who were active at a given time on the basis of their

immigration status. That is, it counts equally for the 1780s Hugh Gaine, who had been printing in New York since 1752, and Thomas Dobson, who arrived in Philadelphia only in 1784.

26. In nearly half of the cases (fifty-four), however, only the date of first activity in the printing trade is known, not the date of immigration.

27. James Horn and Philip D. Morgan, "Settlers and Slaves: European and African Migrations to Early Modern British America," in *The Creation of the British Atlantic World*, ed. Elizabeth Mancke and Carole Shammas (Baltimore: Johns Hopkins University Press, 2005), 19–44; Aaron Fogleman, "Migrations to the Thirteen British North American Colonies, 1700–1775: New Estimates," *Journal of Interdisciplinary History* 22, no. 4 (1992): 691–709; Marianne Wokeck, *Trade in Strangers: The Beginning of Mass Migration to North America* (University Park: Pennsylvania State University Press, 1999). See also Nicholas Canny, ed., *Europeans on the Move: Studies on European Migration, 1500–1800* (Oxford: Clarendon Press, 1994), 76–149; H. Tyler Blethen and Curtis W. Wood Jr., eds., *Ulster and North America: Transatlantic Perspectives on the Scotch-Irish* (Tuscaloosa: University of Alabama Press, 1997).

28. Wokeck, *Trade in Strangers*, 45–46; Horn and Morgan, "Settlers and Slaves," 20–23; Fogleman, "Migration to the Thirteen British North American Colonies," 698–99. On the German printing trade in North America, see A. Gregg Roeber, "German and Dutch Books and Printing," in *CBAW*, 298–314; Robert E. Cazden, *A Social History of the German Book Trade in America to the Civil War* (Columbia, SC: Camden House, 1984), 3–31; Edward W. Hocker, *The Sower Printing House of Colonial Times*, Pennsylvania German Society Proceedings and Addresses, vol. 53 (Norristown, PA: Pennsylvania German Society, 1948).

29. Horn and Morgan, "Settlers and Slaves," 21–22. On Scottish and Irish immigration, see David Dobson, *Scottish Emigration to Colonial America, 1607–1785* (Athens: University of Georgia Press, 1994); Kerby A. Miller, *Emigrants and Exiles: Ireland and the Irish Exodus to North America* (New York: Oxford University Press, 1985).

30. Horn and Morgan, "Settlers and Slaves," 27.

31. Bernard Bailyn, *Voyagers to the West: A Passage in the People of America on the Eve of the Revolution* (New York: Alfred A. Knopf, 1986), 147–60.

32. Bailyn, *Voyagers to the West*, 160.

33. On the spread of printing to rural areas in the early Republic, see Jack Larkin, "'Printing is something every village has in it': Rural Printing and Publishing," in *An Extensive Republic: Print, Culture, and Society in the New Nation, 1790–1840*, ed. Robert A. Gross and Mary Kelley (Chapel Hill: University of North Carolina Press, 2010), 145–60; William J. Gilmore, *Reading Becomes a Necessity of Life: Material and Cultural Life in Rural New England, 1780–1835* (Knoxville: University of Tennessee Press, 1989).

34. John Bidwell, "Printers' Supplies and Capitalization," in *CBAW*, 168–71; Lawrence C. Wroth, *The Colonial Printer* (Charlottesville, VA: Dominion Books, 1964), 82–85, 98–102.

35. Ian Maxted, comp., *The British Book Trades, 1710–1777: An Index of Masters and Apprentices Recorded in the Inland Revenue Registers at the Public Record Office, Kew* (Exeter: published by the author at no. 10, Leighdene Close, 1983), 14. Strahan's career arc resembles those of the subjects of David Hancock's excellent study of Scottish merchants in London during the eighteenth century. Hancock, *Citizens of the World: London Merchants and the Integration of the British Atlantic Community, 1735–1785* (Cambridge: Cambridge University Press, 1995).

36. Wroth, *The Colonial Printer*, 65–67; Bidwell, "Printers' Supplies and Capitalization," 161–81.

37. "Inventory of the Estate of William Rind," *William and Mary Quarterly*, 2nd ser., 17, no. 1 (1937): 53–55. The printing office constituted about 45 percent of his total estate of £272, 5s., 6d.

38. "Quantity and Valuation of the Printing-Office, as taken January 27: 1766. per J Parker," *PBF* 13:60–63.

39. Sheryllyne Haggerty, *The British-Atlantic Trading Community, 1760–1810: Men, Women,*

and the Distribution of Goods (Leiden: Brill, 2006), 109–41; Jennifer J. Baker, *Securing the Commonwealth: Debt, Speculation, and Writing in the Making of Early America* (Baltimore: Johns Hopkins University Press, 2005); Richard L. Bushman, *The Refinement of America: Persons, Houses, Cities* (New York: Vintage Books, 1993).

40. Craig Muldrew, *The Economy of Obligation: The Culture of Credit and Social Relations in Early Modern England* (New York: St. Martin's Press, 1998); Toby L. Ditz, "Secret Selves, Credible Personas: The Problematics of Trust and Public Display in the Writing of Eighteenth-Century Philadelphia Merchants," in *Possible Pasts: Becoming Colonial in Early America*, ed. Robert Blair St. George (Ithaca, NY: Cornell University Press, 2000), 219–42; Hancock, *Citizens of the World*; Doerflinger, *A Vigorous Spirit of Enterprise*, 11–69.

41. Ward L. Miner, *William Goddard: Newspaperman* (Durham, NC: Duke University Press, 1962), 20.

42. Isaiah Thomas to [Zechariah Fowle], March 20, 1772, Isaiah Thomas Papers, AAS.

43. Wroth, *The Colonial Printer*, 80; James Mosley, "The Technologies of Printing," in *The Cambridge History of the Book in Britain*, vol. 5: *1695–1830*, ed. Michael F. Suarez and Michael L. Turner (Cambridge: Cambridge University Press, 2009), 174–82. On the printing presses in use in the eighteenth century, see James Moran, *Printing Presses: History and Development from the Fifteenth Century to Modern Times* (London: Faber and Faber, 1973). I am greatly indebted to Vincent Golden of the American Antiquarian Society for his demonstration and explanation of the printing process, which appear through the chapter.

44. [Benjamin Franklin], "Prices of Printing Work in Philad[a] 1754," Isaiah Thomas Papers, AAS.

45. William Rind in 1768 charged 12s., 6d for his *Virginia Gazette*. William Goddard charged ten shillings for the *Pennsylvania Chronicle*.

46. *Virginia Gazette* (Rind), December 1, 1768.

47. *New-York Journal*, October 16, 1766.

48. "Estimate of the Debts due from the Province of Pennsylvania, 1769," *Pennsylvania Archives*, ed. Samuel Hazard (Philadelphia: printed by Joseph Severns & Co., 1853), ser. 1, IV: 345.

49. See James Parker to Benjamin Franklin, June 11, 1766, *PBF* 13:300–12.

50. On the manufacturing of paper, see Lyman Horace Weeks, *A History of Paper-Manufacturing in the United States, 1690–1916* (New York: Lockwood Trade Journal Co., 1916). By the 1760s, there were about twenty-five paper mills in the American colonies. Bidwell, "Printers' Supplies and Capitalization," 177.

51. David Hall to Henry Unwin, July 25, 1765, David Hall Letter Book, APS.

52. For examples, see Elizabeth Carroll Reilly, *A Dictionary of Colonial American Printers' Ornaments and Illustrations* (Worcester, MA: American Antiquarian Society, 1975), 443–56.

53. On the ideology and morality of business practices in colonial America, see John E. Crowley, *This Sheba, Self: The Conceptualization of Economic Life in Eighteenth-Century America* (Baltimore: Johns Hopkins University Press, 1974).

54. On the long-running dispute over the proprietary system of government, see James H. Hutson, *Pennsylvania Politics, 1746–1770: The Movement for Royal Government and Its Consequences* (Princeton, NJ: Princeton University Press, 1972).

55. Goddard, *The Partnership*, 5–9.

56. "List of William Goddard's Subscribers," February 25, 1767, LCP.

57. William Goddard to John Smith, May 11, 1769, Smith Family Papers, vol. 7, p. 219, LCP in HSP; Bond of William and Mary Katherine Goddard to Benjamin Franklin, December 15, 1769, Benjamin Franklin Papers [coll. 215], HSP.

58. Goddard, *The Partnership*.

59. William Strahan to Benjamin Franklin, August 21, 1772, *PBF* 19:266–67; William Strahan to David Hall, undated, William Strahan Letters [coll. Am.162], HSP.

60. David Hall to William Strahan, January 31, 1767, David Hall Letter Book, APS.

61. Emily Buchnea, "Transatlantic Transformations: Visualizing Change over Time in the Liverpool–New York Trade Network, 1763–1833," *Enterprise & Society* 15, no. 4 (2015): 687–721; Natasha Glaisyer, "Networking: Trade and Exchange in the Eighteenth-Century British Empire," *Historical Journal* 47, no. 2 (2004): 451–76; David Eltis, *The Rise of African Slavery in the Americas* (New York: Cambridge University Press, 2000); David Hancock, "The Trouble with Networks: Managing the Scots' Early-Modern Madeira Trade," *Business History Review* 79, no. 3 (2005): 467–91; Hancock, *Oceans of Wine: Madeira and the Organization of the Atlantic World, 1640–1815* (New Haven, CT: Yale University Press, 2009); Hancock, *Citizens of the World*; John Haggerty and Sheryllynne Haggerty, "Visual Analytics of an Eighteenth-Century Business Network," *Enterprise & Society* 11, no. 1 (2010): 1–23; Marcy Norton, *Sacred Gifts, Profane Pleasures: A History of Tobacco and Chocolate, 1492–1700* (Ithaca, NY: Cornell University Press, 2008); Beiler, *Immigrant and Entrepreneur*.

62. Hancock, "The Trouble with Networks," 478–84.

63. Pamela Walker Laird, *Pull: Networking and Success since Benjamin Franklin* (Cambridge, MA: Harvard University Press, 2006). Charles Wetherell mapped networks for the colonial printing trade and concluded on the basis of Franklin's experience that "the trade was built upon associations and that these were necessary for success." Wetherell, "Brokers of the Word," 19. However, he also argued that growth in the trade during the second half of the eighteenth century made the diffusion of materials through such associations more difficult, a point that this book aims to refute. Laird, *Pull*, 157–81.

64. Ralph Frasca, *Benjamin Franklin's Printing Network: Disseminating Virtue in Early America* (Columbia: University of Missouri Press, 2006), 67–114.

65. *New-York Journal*, July 8, 1773.

66. James N. Green, "English Books and Printing in the Age of Franklin," and Hugh Amory, "The New England Book Trade," in *CBAW*, 248–97, 314–46. On the definition of "bookseller" in eighteenth-century America, see Rosalind Remer, *Printers and Men of Capital: Philadelphia Book Publishers in the New Republic* (Philadelphia: University of Pennsylvania Press, 1996), 2.

67. Marion Barber Stowell, *Early American Almanacs: The Colonial Weekday Bible* (New York: Burt Franklin, 1977), 72–76.

68. Stowell, *Early American Almanacs*, 31. On piracy more generally, see Adrian Johns, *Piracy: The Intellectual Property Wars from Gutenberg to Gates* (Chicago: University of Chicago Press, 2010).

69. Adam Boyd to Isaiah Thomas, December 2, 1769, Isaiah Thomas Papers, AAS.

70. Isaiah Thomas to [Zechariah Fowle], March 20, 1772, Isaiah Thomas Papers, AAS.

71. Hugh Gaine Receipt Book, NYPL, 57; payments to chapel at 45, 97, 193, 234.

72. Jessica Choppin Roney, *Governed by a Spirit of Opposition: The Origins of American Political Practice in Colonial Philadelphia* (Baltimore: Johns Hopkins University Press, 2014). I am very grateful to Jessica for sharing her data on the affiliations of those involved in the book trades in Philadelphia.

73. On the history of the imperial British post office, see Kenneth Ellis, *The Post Office in the Eighteenth Century: A Study in Administrative History* (Oxford: Oxford University Press, 1958); J. C. Hemmeon, *The History of the British Post Office* (Cambridge, MA: Harvard University Press, 1912); Wesley Everett Rich, *The History of the United States Post Office to the Year 1829*, Harvard Economic Studies, vol. 27 (Cambridge, MA: Harvard University Press, 1924); Howard Robinson, *The British Post Office: A History* (Princeton, NJ: Princeton University Press, 1948); William Smith, *The History of the Post Office in British North America, 1639–1870* (Cambridge: Cambridge University Press, 1920; New York: Octagon Books, 1973). See also Ian K. Steele, *The English Atlantic, 1675–1740: An Exploration of Communication and Community* (New York: Oxford University Press, 1986).

74. Smith, *History of the Post Office*, 22–24.

75. *A Collection of the Statutes Now in Force, relating to the Post-Office* (New York: Hugh Gaine, 1774), 105–37; *Georgia Gazette*, May 2, 1765; *Newport Mercury*, May 6, 1765; *Boston Gazette*, May 13, 1765; *South-Carolina Gazette*, October 19, 1765; *Boston Evening Post*, November 25, 1765.

76. Gordon S. Wood, "Conspiracy and the Paranoid Style: Causality and Deceit in the Eighteenth Century," *William and Mary Quarterly*, 3rd ser., 39 (1982): 401–41.

77. Alyssa Zuercher Reichardt, "War for the Interior: Imperial Conflict and the Formation of North American and Transatlantic Communication Infrastructure, 1727–1774" (PhD diss., Yale University, 2017).

78. *A Collection of the Statutes Now in Force*, 26–34, 105–37. Postage rates on letters remained high until a series of postal reforms in the mid-nineteenth century. David M. Henkin, *The Postal Age: The Emergence of Modern Communications in Nineteenth-Century America* (Chicago: University of Chicago Press, 2006).

79. Newspapers in Europe grew out of merchants' desire for the latest commercial news. John J. McCusker, "The Demise of Distance: The Business Press and the Origins of the Information Revolution in the Early Modern Atlantic World," *American Historical Review* 110, no. 2 (2005): 295–321.

80. Hugh Finlay, *Journal kept by Hugh Finlay, surveyor of the post roads on the continent of North America: during his survey of the post offices between Falmouth and Casco Bay, in the province of Massachusetts, and Savannah in Georgia, begun the 13th Septr., 1773 and ended 26th June 1774* (Brooklyn: F. H. Norton, 1867), 28, 32.

81. Finlay, *Journal*, 45–46.

82. Smith, *History of the Post Office*, 25, 51–54. American colonists were not the only ones who avoided paying postage; on the evasion of legal payments in France, see Dena Goodman, *The Republic of Letters: A Cultural History of the French Enlightenment* (Ithaca, NY: Cornell University Press, 1994), 16–20.

83. Steele, *English Atlantic*, 113. See also Richard B. Kielbowicz, *News in the Mail: The Press, Post Office, and Public Information, 1700–1860s* (New York: Greenwood Press, 1989); Wallace B. Eberhard, "Press and Post Office in Eighteenth-Century America: Origins of a Public Policy," in *Newsletters to Newspapers: Eighteenth Century Journalism*, ed. Donovan H. Bond and W. Reynolds McLeod (Morgantown: School of Journalism, West Virginia University, 1977), 145–52.

84. Charles E. Clark, *The Public Prints: The Newspaper in Anglo-American Culture, 1665–1740* (Oxford: Oxford University Press, 1994), 185–89.

85. *Rivington's New-York Gazetteer*, December 30, 1773.

86. See Annabel Patterson, *Milton's Words* (New York: Oxford University Press, 2009), 55–63.

87. Thomas C. Leonard, *The Power of the Press: The Birth of American Political Reporting* (New York: Oxford University Press, 1986), 29–30; J. G. A. Pocock, *The Machiavellian Moment: Florentine Political Thought and the Atlantic Republican Tradition*, 2nd ed. (Princeton, NJ: Princeton University Press, 2003), 467–77. For a modern edition of the letters, see *Cato's Letters; or, Essays on Liberty*, 4 vols. in 2, edited and annotated by Ronald Hamowy (Indianapolis, IN: Liberty Fund, 1995).

88. Robert W. T. Martin, *The Free and Open Press: The Founding of American Democratic Press Liberty* (New York: New York University Press, 2001), 32.

89. David A. Copeland, *The Idea of a Free Press: The Enlightenment and Its Unruly Legacy* (Evanston, IL: Northwestern University Press, 2006), 97–100; Richard Buel Jr., "Freedom of the Press in Revolutionary America: The Evolution of Libertarianism, 1760–1820," in *The Press and the American Revolution*, ed. Bernard Bailyn and John B. Hench (Worcester, MA: Ameri-

can Antiquarian Society, 1980), 59–97; Bernard Bailyn, *The Ideological Origins of the American Revolution*, rev. ed. (Cambridge, MA: Harvard University Press, 1992), 34–93; Pocock, *Machiavellian Moment*, 462–505; Jeffery A. Smith, *Printers and Press Freedom: The Ideology of Early American Journalism* (New York: Oxford University Press, 1988), 24–30; Martin, *The Free and Open Press*, 16–34.

90. Leonard, *The Power of the Press*, 13–32. On early disputes about freedom of the press in the American colonies, see Wm. David Sloan and Julie Hedgepeth Williams, *The Early American Press, 1690–1783* (Westport, CT: Greenwood Press, 1994), 73–95.

91. The newspaper was edited by attorney James Alexander, who had defended the interim governor in a lawsuit filed against him by Cosby. Vincent Buranelli, "Peter Zenger's Editor," *American Quarterly* 7, no. 2 (1955): 174–81. On Zenger, see Copeland, *Idea of a Free Press*, 153–63; Stanley N. Katz, ed., *A Brief Narrative of the Case and Trial of John Peter Zenger, Printer of the New York Weekly Journal* (Cambridge, MA: Belknap Press of Harvard University Press, 1963); William Lowell Putnam, *John Peter Zenger and the Fundamental Freedom* (Jefferson, NC: McFarland, 1997); Thomas, *History of Printing in America*, 2:95–98, 287–93; Patricia U. Bonomi, *A Factious People: Politics and Society in Colonial New York* (New York: Columbia University Press, 1971), 112–20; Paul Finkelman, "Politics, the Press, and the Law: The Trial of John Peter Zenger," in *American Political Trials*, rev. ed., ed. Michal R. Belknap (Westport, CT: Greenwood Press, 1994), 25–44; Michael Warner, *Letters of the Republic: Publication and the Public Sphere in Eighteenth-Century America* (Cambridge, MA: Harvard University Press, 1990), 49–58.

92. Katz, *Brief Narrative*, 9.

93. The narrative here is primarily drawn from Katz, *Brief Narrative*.

94. Botein, "'Meer Mechanics' and an Open Press"; Green, "English Books and Printing," 255–57.

95. *Pennsylvania Gazette*, June 10, 1731 (O.S.). Franklin's argument bears a strong resemblance to the way Habermas described the printer's role in the public sphere, though Habermas did not cite Franklin. See Jürgen Habermas, *The Structural Transformation of the Public Sphere: An Inquiry into a Category of Bourgeois Society*, trans. Thomas Burger (Cambridge, MA: MIT Press, 1989).

96. The argument also resonated because of a tension within colonial American society between the value of work for the "public good" and its value for economic gain. Crowley, *This Sheba, Self*, 2–6.

97. *New-York Gazette; or, the Weekly Post-Boy*, August 20, 1770.

98. *Quebec Gazette*, June 21, 1764.

Chapter 2 · A Trade under Threat: Printers and the Stamp Act Crisis

1. Benjamin Franklin to John Ross, February 14, 1765, *PBF* 12:68. See also Benjamin Franklin to David Hall, February 14, 1765, *PBF* 12:65–67.

2. Thomas P. Slaughter, "The Tax Man Cometh: Ideological Opposition to Internal Taxes, 1760–1790," *William and Mary Quarterly*, 3rd ser., 41 (1984): 566–91; Fred Anderson, *The Crucible of War: The Seven Years' War and the Fate of Empire in British North America, 1754–1766* (New York: Vintage, 2001), 641–708.

3. Gary B. Nash, *The Urban Crucible: Social Change, Political Consciousness, and the Origins of the American Revolution* (Cambridge, MA: Harvard University Press, 1979), 295–96; Pauline Maier, *From Resistance to Revolution: Colonial Radicals and the Development of American Opposition to Britain, 1765–1776* (New York: Alfred A. Knopf, 1972), 51–76.

4. Edmund S. Morgan and Helen M. Morgan, *The Stamp Act Crisis: Prologue to Revolution* (Chapel Hill: IEAHC, University of North Carolina Press, 1962; repr., 1995).

5. Stephen Botein, "'Meer Mechanics' and an Open Press: The Business and Political Strategies of Colonial American Printers," *Perspectives in American History* 9 (1975): 127–225; Botein,

"Printers and the American Revolution," in *The Press and the American Revolution*, ed. Bernard Bailyn and John B. Hench (Worcester, MA: American Antiquarian Society, 1980), 11–54; Arthur M. Schlesinger, *Prelude to Independence: The Newspaper War on Britain, 1764–1776* (New York: Alfred A. Knopf, 1957); Philip G. Davidson, *Propaganda and the American Revolution, 1763–1783* (Chapel Hill: University of North Carolina Press, 1941).

6. P. D. G. Thomas, *British Politics and the Stamp Act Crisis: The First Phase of the American Revolution, 1763–1767* (Oxford: Clarendon Press, 1975); Eliga H. Gould, *The Persistence of Empire: British Political Culture in the Age of the American Revolution* (Chapel Hill: OIEAHC, University of North Carolina Press, 2000), 110–22; Fred Anderson, *The Crucible of War: The Seven Years' War and the Fate of Empire in British North America, 1754–1766* (New York: Alfred A. Knopf, 2000), 641–708; Morgan and Morgan, *Stamp Act Crisis*.

7. Anderson, *Crucible of War*, 581–87.

8. Anderson, *Crucible of War*, 572–80.

9. See, for example, an essay by "A Lover of Pennsylvania," *Pennsylvania Journal*, June 28, 1764; reprinted in the *Newport Mercury*, July 16, 1764.

10. James Otis, *The Rights of the British Colonies Asserted and Proved* (Boston: Edes and Gill, 1764), Early American Imprints, 1st ser., no. 9773. For the advertisements, see *Boston Gazette*, July 23 and August 13, 1764; *Boston Post-Boy*, July 23, July 30, August 6, and August 13, 1764; *Boston Evening Post*, July 23, August 6, and August 13, 1764; *Boston News-Letter*, July 26 and August 9, 1764.

11. Morgan and Morgan, *Stamp Act Crisis*, 54–74; Anderson, *Crucible of War*, 610–14.

12. *Anno Regni Georgii III. . . . QUINTO. An Act for granting and applying certain Stamp Duties, and other Duties, in the British Colonies and Plantations in America* (London: Mark Baskett, 1765), 279–97. For a categorical breakdown of the duties, see Adolph Koeppel, *The Stamps That Caused the American Revolution: The Stamps of the 1765 British Stamp Act for America* (Manhasset, NY: Town of North Hempstead American Revolution Bicentennial Commission, 1976), 59–67.

13. Any copies left unsold could then be returned to the commissioners for credit on the printer's account. *An Act for granting and applying certain Stamp Duties*, 297; Koeppel, *The Stamps That Caused the American Revolution*, 69–72.

14. Anderson, *Crucible of War*, 645.

15. John Bidwell, "Printers' Supplies and Capitalization," in *CBAW*, 178; Michael Harris, *London Newspapers in the Age of Walpole: A Study of the Origins of the Modern English Press* (Rutherford, NJ: Fairleigh Dickinson University, 1987), 19–30; Jeremy Black, *The English Press in the Eighteenth Century* (London: Croom Helm, 1987), 106–7.

16. Mack Thompson, "Massachusetts and New York Stamp Acts," *William and Mary Quarterly*, 3rd ser., 26, no. 1 (1969): 253–58; Andrew Jackson O'Shaughnessy, *An Empire Divided: The American Revolution and the British Caribbean* (Philadelphia: University of Pennsylvania Press, 2000), 96. On the "mixed response" to these stamp acts among printers in the Franklin network, see Ralph Frasca, *Benjamin Franklin's Printing Network: Disseminating Virtue in Early America* (Columbia: University of Missouri Press, 2006), 138–39.

17. *The Colonial Laws of New York, from the Year 1664 to the Revolution . . .*, 4 vols. (Albany: James B. Lyon, State Printer, 1894), 4:110–16.

18. Edmund Morgan and Helen Morgan suggested that "the lower classes probably had little to lose directly by the Stamp Act." *Stamp Act Crisis*, 194.

19. Carl Robert Keyes, "Early American Advertising: Marketing and Consumer Culture in Eighteenth-Century Philadelphia" (PhD diss., Johns Hopkins University, 2007), especially chap. 2.

20. David Hall to William Strahan, May 19, 1765, David Hall Letter Book, APS.

21. David Hall to Benjamin Franklin, June 20, 1765, *PBF* 12:188–89.

22. David Hall to William Strahan, May 19, 1765, David Hall Letter Book, APS.
23. William Strahan to David Hall, July 8, 1765, David Hall Papers, APS.
24. Benjamin Franklin to David Hall, June 8, 1765, *PBF* 12:171–72.
25. William Strahan to David Hall, August 23, 1765, David Hall Papers, APS.
26. Historians have agreed with his assessment. David Waldstreicher wrote that Franklin and Strahan demonstrated "little understanding of how strapped colonists, and especially printers, were for cash" in advising Hall to operate on a "ready money" basis. David Waldstreicher, *Runaway America: Benjamin Franklin, Slavery, and the American Revolution* (New York: Hill and Wang, 2004), 166.
27. *Pennsylvania Journal*, June 13, 1765; *New-York Gazette* (Holt), June 20, 1765. Editions were also printed by Richard and Samuel Draper and Green and Russell (Boston, 1765); Edes and Gill (Boston, 1765); Timothy Green (New London, 1765); and James Parker ([Woodbridge, NJ], 1765).
28. See, for example, *Boston Gazette*, April 8, 1765; *New-York Gazette* (Holt), April 11, 1765.
29. *Ames's Almanack revived and improved: Or, An Astronomical Diary, for the Year of our Lord CHRIST 1766* (Boston: R. & S. Draper, Edes & Gill, Green & Russell, T. & J. Fleet, S. Hall, 1765).
30. Matthew J. Shaw, "Keeping Time in the Age of Franklin: Almanacs and the Atlantic World," *Printing History*, n.s., no. 2 (2007): 17–37; Allan R. Raymond, "To Reach Men's Minds: Almanacs and the American Revolution, 1760–1777," *New England Quarterly* 51, no. 3 (1978): 370–95.
31. *Poor Richard improved: Being an Almanack and Ephemeris . . . for the Year of our Lord 1766* (Philadelphia: B. Franklin, and D. Hall, 1765).
32. *Boston Gazette*, July 29, 1765. *Poor Richard* was first advertised in the *Pennsylvania Gazette* on July 11, 1765.
33. *Pennsylvania Journal*, August 1, 1765; *Providence Gazette, and Country Journal*, May 4, 1765; *New-York Gazette* (Holt), September 19, 1765.
34. *Pennsylvania Journal*, July 11, 25, and August 8, 1765.
35. *The Gentleman and Citizen's Pocket-Almanack, for the Year 1766* (Philadelphia: Andrew Steuart, 1765); *The Pennsylvania Pocket Almanack, For the Year 1766* (Philadelphia: William Bradford, 1765).
36. Botein, "'Meer Mechanics' and an Open Press."
37. *An Act for granting and applying certain Stamp Duties*, 292–97.
38. On reaction to the Stamp Act in the Caribbean, see O'Shaughnessy, *An Empire Divided*, 81–108.
39. *Barbados Mercury*, April 19, 1766, LCP; O'Shaughnessy, *An Empire Divided*, 93.
40. *Boston Gazette*, February 3, 1766; the article also appeared in the *Pennsylvania Gazette*, February 13, 1766.
41. *Providence Gazette, and Country Journal*, May 4, 1765. Goddard published a special unnumbered issue of the *Providence Gazette* on August 24, 1765, but by the time regular publication resumed in 1766, he had turned the office over to his mother, Sarah Updike Goddard, assisted by his onetime apprentice, John Carter.
42. James Parker to Benjamin Franklin, April 25, 1765, *PBF* 12:111.
43. Beverly McAnear, "James Parker versus William Weyman," *Proceedings of the New Jersey Historical Society* 59, no. 1 (1941): 1–23; McAnear, "James Parker versus John Holt," *Proceedings of the New Jersey Historical Society* 59, no. 2 (1941): 77–95; and no. 3: 198–211.
44. *Maryland Gazette*, October 10, 1765. Green published three "supplements" to the final number of the *Maryland Gazette*, which came out on the regular Thursday publication date through October 31. He eventually resumed publication, despite his rhetoric. He published an extraordinary edition on December 10, 1765, and then began weekly publication again in February 1766.

45. *Boston Evening-Post*, October 28, 1765; *Boston Gazette*, November 4, 1765.

46. Hall's sheets were published November 7 and 14, 1765.

47. [William Goddard] to Isaiah Thomas, n.d. [after 1810], Isaiah Thomas Papers, AAS.

48. Morgan and Morgan, *Stamp Act Crisis*, 102.

49. Alfred L. Lorenz, *Hugh Gaine: A Colonial Printer-Editor's Odyssey to Loyalism* (Carbondale: Southern Illinois University Press, 1972), 38–39.

50. For information on the publication record of these newspapers, see Clarence S. Brigham, *History and Bibiliography of American Newspapers, 1690–1820*, 2 vols. (Worcester, MA: American Antiquarian Society, 1947), 1:290–331.

51. *New-York Mercury*, October 21, 1765; *Pennsylvania Journal*, September 19, 1765; *Boston Gazette*, October 21, 1765.

52. See, for example, [New York Sons of Liberty] to William Bradford, February 13, 1766, Pennsylvania Stamp Act and Non-Importation Resolutions Collection, 1765–1775, APS; Philadelphia Sons of Liberty to New York Sons of Liberty, February 15, 1766, and William Bradford to John Lamb, Isaac Sears, Robinson, William Willey, and Gershom Mott, February 15, 1766, John Lamb Papers, N-YHS.

53. Morgan and Morgan, *Stamp Act Crisis*, 128; Ward L. Miner, *William Goddard, Newspaperman* (Durham, NC: Duke University Press, 1962), 49.

54. McAnear, "James Parker versus John Holt [I]," 89.

55. Charles E. Clark, *The Public Prints: The Newspaper in Anglo-American Culture, 1665–1740* (New York: Oxford University Press, 1994), 215–21.

56. *Boston Evening-Post*, August 19, 1765.

57. *New-Hampshire Gazette*, August 23, 1765; *Newport Mercury*, August 26, 1765; *New-York Mercury*, August 26, 1765; *New-York Gazette*, August 29, 1765; *Pennsylvania Journal*, August 29, 1765; *Pennsylvania Gazette*, August 29, 1765.

58. With the passage of the Stamp Act, George Grenville thought that Americans might accept the tax more readily if it was collected by Americans, and Franklin decided that since he could not prevent the Stamp Act from taking place, he might at least secure some of its profits for his American friends. Franklin did nominate his friend John Hughes as Pennsylvania stamp collector. Gordon S. Wood, *The Americanization of Benjamin Franklin* (New York: Penguin Books, 2004), 112.

59. *South-Carolina Gazette*, October 19, 1765.

60. *South-Carolina Gazette*, October 31, 1765. On Timothy's reaction to the Stamp Act, see Hennig Cohen, *The South Carolina Gazette, 1732–1775* (Columbia: University of South Carolina Press, 1953), 13; Susan Macall Allen, "The Impact of the Stamp Act of 1765 on Colonial American Printers: Threat or Bonanza?" (PhD diss., UCLA, 1996), 69.

61. Peter Timothy to Benjamin Franklin, September 3, 1768, *PBF* 15:199–203; *South-Carolina Gazette*, October 19, 1765.

62. John Hampden was a noted seventeenth-century English republican. *Oxford Dictionary of National Biography*, s.v. "Hampden, John," http://www.oxforddnb.com.

63. "Remarkable *Occurrences*," [*Pennsylvania Gazette*], November 14, 1765. See also "Liberty and Property, and NO STAMPS" [*Connecticut Gazette* (New London: Timothy Green)], November 15, 1765.

64. McAnear, "James Parker versus John Holt."

65. Francis Bernard to the Board of Trade, November 30, 1765, American Papers of the Second Earl of Dartmouth, 1765–1782, Staffordshire Record Office, D(W)1778/II/116, viewed at DLAR.

66. Sworn deposition of Normand Tolmie, March 20, 1766; see also the depositions of Isaac Sears, Abraham Montany, Thomas Ivers, Willm. Wiley, and James Dunscomb, March 20, 1766, John Lamb Papers [MS 361], N-YHS. Dennys de Berdt, a colonial agent, made a similar refer-

ence to the 1745 Jacobite uprising, suggesting it to Lord Dartmouth as an example against taxing the colonists. De Berdt to Dartmouth, September 5, 1765, Dartmouth American Papers, D(W) 1778/II/83, viewed at DLAR.

67. *New-York Mercury*, March 24, 1766; *Pennsylvania Gazette*, March 27, 1766; *Boston Gazette*, April 7, 1766.

68. E. P. Thompson, *Customs in Common: Studies in Traditional Popular Culture* (New York: New Press, 1993), 16–96; Maier, *From Resistance to Revolution*; Dirk Hoerder, *Crowd Action in Revolutionary Massachusetts, 1765–1780*, Studies in Social Discontinuity (New York: Academic Press, 1977), 40–84; Thomas P. Slaughter, "Crowds in Eighteenth-Century America: Reflections and New Directions," *Pennsylvania Magazine of History and Biography* 115, no. 1 (1991): 3–34.

69. Michael Meranze, *Laboratories of Virtue: Punishment, Revolution, and Authority in Philadelphia, 1760–1835* (Chapel Hill: OIEAHC, University of North Carolina Press, 1996), 19–54; Mark Fearnow, "American Colonial Disturbances as Political Theatre," *Theatre Survey* 33 (1992): 53–64. Paul Friedland has made similar arguments about the uses of execution as a form of state legitimation in France. See "Beyond Deterrence: Cadavers, Effigies, Animals and the Logic of Executions in Pre-modern France," *Historical Reflections/Réflexions Historiques* 29, no. 2 (2003), 295–317.

70. For an account of the riots in Boston, see Morgan and Morgan, *Stamp Act Crisis*, 125–49.

71. Hoerder, *Crowd Action in Revolutionary Massachusetts*, 90.

72. *Boston Evening-Post*, September 2, 1765.

73. *Maryland Gazette*, September 5, 1765; *Pennsylvania Gazette*, September 19, 1765; *New-York Gazette*, September 19, 1765; *Boston Evening-Post*, September 23, 1765; *Boston Gazette*, September 23, 1765.

74. On the literary and artistic aspects of the Stamp Act protests, see Kenneth Silverman, *A Cultural History of the American Revolution: Painting, Music, Literature, and the Theatre in the Colonies and the United States from the Treaty of Paris to the Inauguration of George Washington, 1763–1789* (New York: Thomas Y. Crowell, 1976), 72–99.

75. *New-York Gazette*, August 29, 1765; *New York Mercury*, September 2, 1765.

76. William Pencak, "Play as Prelude to Revolution: Boston, 1765–1776," in *Riot and Revelry in Early America*, ed. William Pencak, Matthew Dennis, and Simon P. Newman (University Park: Pennsylvania State University Press, 2002), 125–55.

77. *Boston Evening-Post*, September 2, 1765; *Boston Gazette*, September 2, 1765. On the role of the gentry in the Newport Stamp Act riots, see Allen Mansfield Thomas, "'Circumstances not Principles': Elite Control of the Newport Stamp Act Riots," *Newport History* 67, no. 3 (1995): 129–43.

78. Morgan and Morgan, *Stamp Act Crisis*, 163–64; *Boston Gazette*, August 19, 1765; *Providence Gazette*, August 24, 1765; *Newport Mercury*, August 26, 1765.

79. The *Newport Mercury* republished both pieces in the same issue. *Boston Evening-Post*, August 19, 1765; *New-Hampshire Gazette*, August 23, 1765; *Newport Mercury*, August 26, 1765; *New-York Gazette*, August 29, 1765; *New-York Mercury*, August 29, 1765; *Pennsylvania Journal*, August 29, 1765; *Pennsylvania Gazette*, August 29, 1765.

80. Maier, *From Resistance to Revolution*, 77–79.

81. *New-York Gazette*, August 22, 1765.

82. *Massachusetts Gazette. And Boston News-Letter*, August 29, 1765.

83. *Constitutional Courant* (Woodbridge, NJ), September 21, 1765. On identifying Goddard as the printer, see Miner, *William Goddard*, 50–1; Ralph Frasca, "'At the Sign of the Bribe Refused': The *Constitutional Courant* and the Stamp Tax, 1765," *New Jersey History* 107 (1989): 21–39. The newspaper even reached the desk of Lord Dartmouth, who at the time was head of the Board of Trade. See Dartmouth American Papers, D(W) 1778/II/87, viewed at DLAR.

84. Annabel Patterson, *Marvell: The Writer in Public Life* (Harlow, England: Longman, 2000); Patterson, introduction to *The Prose Works of Andrew Marvell,* vol. 1 (New Haven, CT: Yale University Press, 2003), xi–xliii. In addition to his poetry, for which he may be better known today, Marvell published several treatises decrying corruption in government. See J. G. A. Pocock, *The Machiavellian Moment: Florentine Political Thought and the Atlantic Republican Tradition* (Princeton, NJ: Princeton University Press, 1975), 408–9; Annabel Patterson, *Early Modern Liberalism* (Cambridge: Cambridge University Press, 1997). Frasca speculated that Goddard himself wrote the essays. See "'At the Sign of the Bribe Refused.'"

85. *New-York Mercury,* August 12, 1765. The letter first appeared in several London newspapers, including the *London Chronicle,* April 23, 1765, *St. James Chronicle,* April 23, 1765, and *Public Advertiser,* April 24, 1765. It was reprinted in America in the *Newport Mercury,* August 26, 1765.

86. *Constitutional Courant* (Woodbridge, NJ), September 21, 1765.

87. William Goddard to Isaiah Thomas, April 22, 1811, Isaiah Thomas Papers, AAS.

88. *Newport Mercury,* October 7, 1765; *New-London Gazette,* October 25, 1765.

89. *Boston Evening-Post,* October 7, 1765. Marvel's address and one of the essays also appeared in the *Connecticut Gazette,* October 4, 1765.

90. *Boston News-Letter,* January 16, 1766; *Boston Post-Boy,* January 20, 1766.

91. *New-Hampshire Gazette,* September 26, 1766. The Marvell quotation, in which he described the press satirically as "that *villainous* Engine," appeared in *The rehearsal transpros'd, or, Animadversions upon a late book intituled, A preface, shewing what grounds there are of fears and jealousies of popery* (London: Printed by J.D. for the assigns of John Calvin and Theodore Beza . . . and sould by N. Ponder . . ., 1672), 4–5.

92. *Boston Gazette,* December 16, 1765.

93. *Boston Gazette,* December 23, 1765. For an example of a similar situation among tea commissioners, see Toby L. Ditz, "Secret Selves, Credible Personas: The Problematics of Trust and Public Display in the Writing of Eighteenth-Century Philadelphia Merchants," in *Possible Pasts: Becoming Colonial in Early America,* ed. Robert Blair St. George (Ithaca, NY: Cornell University Press, 2000), 233–41.

94. *Boston Gazette,* February 24, 1766. Robert Blair St. George has discussed the image as one of a number highlighting effigies in Revolutionary America. See Robert Blair St. George, *Conversing by Signs: Poetics of Implication in Colonial New England Culture* (Chapel Hill: University of North Carolina Press, 1998), 252–54.

95. Keyes, "Early American Advertising," chap. 2.

96. *Boston Gazette,* February 24, 1766.

97. See, for example, Mal. 3:5; Isa. 1:17; Exod. 22:22; Zech. 7:10.

98. Nicole Eustace, *Passion Is the Gale: Emotion, Power, and the Coming of the American Revolution* (Chapel Hill: OIEAHC, University of North Carolina Press, 2008), 394–407. As T. H. Breen has shown, some towns, Boston in particular, worked to simplify the mourning process during the 1750s and 1760s out of a belief that funerals had become too commodified and extravagant. T. H. Breen, *The Marketplace of Revolution: How Consumer Politics Shaped American Independence* (New York: Oxford University Press, 2004), 213–17.

99. Thomas, *History of Printing in America,* 1:370–74. For an alternative account, see John Bartlet Brebner, *The Neutral Yankees of Nova Scotia: A Marginal Colony during the Revolutionary Years* (New York: Columbia University Press, 1937), 158–62.

100. Morgan and Morgan, *Stamp Act Crisis,* 165–86.

101. William Strahan to David Hall, April 7, 1766, Strahan Letters [coll. Am.162], HSP. *The examination of Doctor Benjamin Franklin, before an august assembly, relating to the repeal of the stamp-act, &c.* (Philadelphia: Hall and Sellers, 1766). James Parker and Edes and Gill re-printed the pamphlet.

102. See, for example, Nathaniel Ames, *An Astronomical Diary; or, Almanack for the Year of our Lord Christ 1767* (Boston: William M'Alpine, 1766); [Andrew Aguecheek], *The Universal American Almanack, or Yearly Magazine. . . . for the Year of our Lord 1767* (Philadelphia: Andrew Steuart, 1766).

Chapter 3 · The Business of Protest: Printing against Empire

1. Peter Thompson, *Rum Punch and Revolution: Taverngoing & Public Life in Eighteenth-Century Philadelphia* (Chapel Hill: OIEAHC, University of North Carolina Press, 1999), 10–11; David W. Conroy, *In Public Houses: Drink and the Revolution in Authority in Colonial Massachusetts* (Chapel Hill: IEAHC, University of North Carolina Press, 1995).

2. Hugh Gaine receipt book, NYPL; Vincent DiGirolamo, "In Franklin's Footsteps: News Carriers and Postboys in the Revolution and Early Republic," in *Children and Youth in a New Nation*, ed. James Marten (New York: New York University Press, 2009), 48–66.

3. Nathanael Low, *An Astronomical Diary; Or, Almanack, for the Year of Christian Æra, 1771* (Boston: Kneeland and Adams, 1770); Allan R. Raymond, "To Reach Men's Minds: Almanacs and the American Revolution, 1760–1777," *New England Quarterly* 51, no. 3 (1978): 370–95.

4. William Goddard, *The Partnership: Or the History of the Rise and Progress of the Pennsylvania Chronicle, &c. Wherein the Conduct of* JOSEPH GALLOWAY, *Esq; Speaker of the Honourable House of Representatives of the Province of* Pennsylvania, *Mr.* THOMAS WHARTON, *sen. and* their Man BENJAMIN TOWNE, *my late Partners with my own, is properly delineated, and their Calumnies against me fully refuted* (Philadelphia: William Goddard, 1770), Early American Imprints, 1st ser., no. 11669.

5. Joseph S. Tiedemann, *Reluctant Revolutionaries: New York City and the Road to Independence, 1763–1776* (Ithaca, NY: Cornell University Press, 1997), 70.

6. John Adams, entry of September 3, 1769, *Diary and Autobiography of John Adams*, 4 vols., ed. L. H. Butterfield (Cambridge, MA: Belknap Press of Harvard University Press, 1961), 1:343.

7. Robert Middlekauff, *The Glorious Cause: The American Revolution, 1763–1789* (New York: Oxford University Press, 1982), 150.

8. T. H. Breen, *The Marketplace of Revolution: How Consumer Politics Shaped American Independence* (New York: Oxford University Press, 2004), 235–53. Thomas Doerflinger argues that in Philadelphia, merchants gladly went along with the 1767 boycotts because of excess inventory in their warehouses. Once the glut had cleared, at the sign of any concessions by the British, Philadelphia merchants pushed to end the agreements so that they could resume trade. Doerflinger, *A Vigorous Spirit of Enterprise: Merchants and Economic Development in Revolutionary Philadelphia* (Chapel Hill: IEAHC, University of North Carolina Press, 1986), 180–96.

9. Bernard Bailyn, *The Ideological Origins of the American Revolution*, rev. ed. (Cambridge, MA: Harvard University Press, 1992).

10. *Boston Evening-Post*, March 22, 1773. According to the America's Historical Imprints database, the pamphlet went through four editions total. Benjamin Church, *An Oration; Delivered March 5th, 1773, at the Request of the Inhabitants of Boston* (Boston: [J. Greenleaf], 1773); (Boston: Edes & Gill, 1773); 4th ed. (Boston: J. Greenleaf, 1773); 4th ed. (Salem: Samuel and Ebenezer Hall, 1773).

11. Lindsay O'Neill, *The Opened Letter: Networking in the Early Modern British World* (Philadelphia: University of Pennsylvania Press, 2014); Konstantin Dierks, *In My Power: Letter Writing and Communications in Early America* (Philadelphia: University of Pennsylvania Press, 2009); Sarah M. S. Pearsall, *Atlantic Families: Lives and Letters in the Later Eighteenth Century* (New York: Oxford University Press, 2008); Eve Tavor Bannet, *Empire of Letters: Letter Manuals and Transatlantic Correspondence, 1688–1820* (Cambridge: Cambridge University Press, 2005); William Merrill Decker, *Epistolary Practices: Letter Writing in America before Telecommunications* (Chapel Hill: University of North Carolina Press, 1998).

12. Jane Calvert, *Quaker Constitutionalism and the Political Thought of John Dickinson* (Cambridge: Cambridge University Press, 2009), 177–203; James H. Hutson, *Pennsylvania Politics, 1746–1770: The Movement for Royal Government and Its Consequences* (Princeton, NJ: Princeton University Press, 1972).

13. *An Address to the Committee of Correspondence in Barbados. . . . By a North-American* (Philadelphia: William Bradford, 1766); Calvert, *Quaker Constitutionalism*, 208–11; Hutson, *Pennsylvania Politics, 1746–1770*, 156–59; Milton E. Flower, *John Dickinson: Conservative Revolutionary* (Charlottesville: Published for the Friends of the John Dickinson Mansion by the University Press of Virginia, 1983).

14. Jane E. Calvert, "Liberty without Tumult: Understanding the Politics of John Dickinson," *Pennsylvania Magazine of History and Biography* 131, no. 3 (2007): 233–62.

15. On Dickinson's arguments and general political philosophy, see Calvert, *Quaker Constitutionalism*; Carl F. Kaestle, "The Public Reaction to John Dickinson's *Farmer's Letters*," *Proceedings of the American Antiquarian Society* 78, no. 2 (1969): 323–53; Pierre Marambaud, "Dickinson's *Letters From a Farmer in Pennsylvania* as Political Discourse: Ideology, Imagery, and Rhetoric," *Early American Literature* 12 (1977): 63–72; A. Owen Aldridge, "Paine and Dickinson," *Early American Literature* 11 (1976): 125–38.

16. *Pennsylvania Gazette*, January 14, 1768.

17. *Pennsylvania Gazette*, January 21, 1768.

18. Kaestle, "Public Reaction," 331. On republicanism, see Bailyn, *Ideological Origins*; Gordon S. Wood, *The Creation of the American Republic, 1776–1787* (Chapel Hill: OIEAHC, University of North Carolina Press, 1998).

19. The first letter in the series was pointedly dated November 5, "*The day of King* WILLIAM *the Third's landing*," *Pennsylvania Gazette*, December 3, 1767.

20. *Pennsylvania Gazette*, February 18, 1768.

21. John Dickinson to James Otis, January 25, 1768, *Warren-Adams Letters, Being chiefly a correspondence among John Adams, Samuel Adams, and James Warren*, vol. 1, Collections of the Massachusetts Historical Society, vols. 72–73 (Boston: Massachusetts Historical Society, 1917), 5.

22. On time lags, see Allan R. Pred, "Urban Systems Development and the Long-Distance Flow of Information through Preelectronic U.S. Newspapers," *Economic Geography* 47 (1971): 498–524.

23. Why this was so is unclear at this point, but bears further research.

24. Boston: Edes and Gill; Boston: Mein and Fleeming; Philadelphia: Hall and Sellers (2 eds.); New York: John Holt; Philadelphia: William Bradford (3rd ed.); Williamsburg: William Rind. See Thomas R. Adams, *American Independence: The Growth of an Idea* (Providence, RI: Brown University Press, 1965), 37–41.

25. *Boston News-Letter*, March 24, 1768; *Boston Evening-Post*, March 28, 1768; *Boston Gazette*, March 28, 1768; *Boston Post-Boy*, March 28, 1768; *Pennsylvania Chronicle*, April 4, 1768; *Pennsylvania Gazette*, April 7, 1768; *New-York Gazette, and Weekly Mercury*, April 11, 1768; *Maryland Gazette*, April 14, 1768.

26. "An Address from the Moderator and Freemen of the Town of Providence," *Providence Gazette*, June 25, 1768; *Pennsylvania Chronicle*, July 4, 1768; *New-York Journal*, July 9, 1768; *Georgia Gazette*, September 14, 1768. "An ADDRESS of the GRAND JURY for the County of Cecil," *Pennsylvania Gazette*, September 1, 1768; *Pennsylvania Chronicle*, September 5, 1768.

27. Goddard, *The Partnership*.

28. *Letters from a Farmer in Pennsylvania, to the Inhabitants of the British Colonies* (London: printed for J. Almon, opposite Burlington-house, Piccadilly, 1768).

29. *Letters from a Farmer* (London: J. Almon, 1768), iii.

30. Adams, *American Independence*, 39–41.

31. *Gentleman's Magazine* (London: Printed at St. John's Gate, for David Henry), June 1768, review no. 34. The review was reprinted in the following American newspapers: *Pennsylvania Chronicle*, September 12, 1768; *New-York Gazette*, September 19, 1768; *New-York Journal*, September 22, 1768; *Boston Chronicle*, September 26, 1768; *Boston Evening-Post*, September 26, 1768.

32. *Pennsylvania Gazette*, January 26, 1769; William Strahan to David Hall, November 10, 1768, William Strahan Letters [coll. Am.162], HSP.

33. Edward Connery Lathem, *Chronological Tables of American Newspapers, 1690–1820, Being a Tabular Guide to Holdings of Newspapers Published in America through the Year 1820* (Barre, MA: American Antiquarian Society and Barre Publishers, 1972), 6–9; Mary Ann Yodelis, *Who Paid the Piper? Publishing Economics in Boston, 1763–1775*, Journalism Monographs, no. 38 (Lexington, KY: Association for Education in Journalism, 1975).

34. John E. Alden, "John Mein, Scourge of Patriots," *Publications of the Colonial Society of Massachusetts* 34 (1942): 572–99.

35. Yodelis, *Who Paid the Piper?*, 27–28.

36. Thomas Hutchinson to [?], February 3, 1771, quoted in Yodelis, *Who Paid the Piper?*, 27.

37. Eric Hinderaker, *Boston's Massacre* (Cambridge, MA: Belknap Press of Harvard University Press, 2017), 134–39; Richard Archer, *As If an Enemy's Country: The British Occupation of Boston and the Origins of Revolution*, Pivotal Moments in American History (New York: Oxford University Press, 2010), 126–27.

38. *Massachusetts Gazette and Boston News-Letter*, March 8, 1770; *Boston Chronicle*, March 8, 1770.

39. *Boston News-Letter*, March 8, 1770.

40. *Newport Mercury*, March 12, 1770; *New-York Gazette, or Weekly Post Boy*, March 19, 1770; *New-York Gazette, and Weekly Mercury*, March 19, 1770.

41. *Boston Gazette*, March 12, 1770.

42. Hiller B. Zobel, *The Boston Massacre* (New York: W. W. Norton, 1970), 215.

43. *Boston Evening Post*, March 12, 1770; *Essex Gazette* (Salem), March 13, 1770; *Connecticut Journal*, March 16, 1770; *Connecticut Gazette* (New London), March 16, 1770; *Providence Gazette*, March 17, 1770; *Pennsylvania Gazette*, March 22, 1770; *Virginia Gazette* (Williamsburg: Purdie & Dixon), April 5, 1770; *Georgia Gazette*, April 11, 1770.

44. Peter Messer, "'A Scene of Villainy Acted by a Dirty Banditti, as Must Astonish the Public': The Creation of the Boston Massacre," *New England Quarterly* 90, no. 4 (December 2017): 502–39.

45. Hinderaker, *Boston's Massacre*, 221–34.

46. Richard D. Brown, *Revolutionary Politics in Massachusetts: The Boston Committee of Correspondence and the Towns* (Cambridge, MA: Harvard University Press, 1970).

47. Isaiah Thomas, *The History of Printing in America, with a Biography of Printers, and an Account of Newspapers*, 2 vols. (Worcester, MA: Isaiah Thomas, Jr., 1810), 1:391–92; Francis G. Walett, "Joseph Greenleaf: Abington Patriot and Propagandist," in *Old Abington in the American Revolution: A Bicentennial Publication*, ed. Dyer Memorial Library Trustees (Abington, MA: Dyer Memorial Library, 1976), 7–17. One biographer of Dr. Joseph Warren suspects that he and not Greenleaf was actually Mucius Scævola. See Samuel A. Forman, "If I had Power to Restrain the Press, I Should Have No Inclination to Hinder *Mucius*," *Dr. Joseph Warren on the Web* (blog), May 1, 2012, http://www.drjosephwarren.com/2012/05/if-i-had-power-to-restrain-the-press-i-should-have-no-inclination-to-hinder-mucius/.

48. Samuel Adams, "Verses satirizing the prominent men of the revolutionary party in Boston," December 4, 1772, Samuel Adams Papers, NYPL.

49. Isaiah Thomas, *Three Autobiographical Fragments; Now First Published upon the 150th Anniversary of the Founding of the American Antiquarian Society, October 24, 1812* (Worcester, MA: American Antiquarian Society, 1962), 13.

50. November 30, 1772, Minute Book I, 4–5, BCC Papers. The committee also voted that Boston's selectmen should receive six copies of the pamphlet, its representatives twelve, that all clergy in the town receive a copy, and that John Hancock receive six on top of those he received as a selectman and representative.

51. Richard D. Brown, *Knowledge Is Power: The Diffusion of Information in Early America, 1700–1865* (New York: Oxford University Press, 1989), 65–82.

52. *The votes and proceedings of the freeholders and other inhabitants of the town of Boston, in town meeting assembled, according to law. (Published by order of the town.)* (Boston: Printed by Edes and Gill, in Queen-Street, and T. and J. Fleet, in Cornhill, [1772]); Brown, *Revolutionary Politics in Massachusetts*, 68–80.

53. *Votes and Proceedings*, 30–36, quote on 33.

54. *Votes and Proceedings*, 34.

55. January 12, 1773, Minute Book I, 49, BCC Papers.

56. December 1, 1772, Minute Book I, 6–8, BCC Papers.

57. *Massachusetts Spy*, December 4, 1772; reprinted in the *Pennsylvania Journal*, December 16, 1772.

58. *Massachusetts Gazette, and Boston Weekly News-Letter*, November 26, 1772. The *Boston Gazette* published the material on November 23, 1772.

59. Minute Book I, 6–90; II, 91–180, BCC Papers.

60. January 19, 1773, Minute Book II, 91, BCC Papers.

61. January 26, 1773, Minute Book II, 180; February 2, 1773, Minute Book III, 181; both in BCC Papers.

62. Darius Sessions, Stephen Hopkins, John Cole, and Moses Brown to Samuel Adams, December 25, 1772, Samuel Adams Papers, NYPL. On the burning of the *Gaspée*, see Middlekauff, *The Glorious Cause*, 213–14.

63. *Massachusetts Spy*, December 31, 1772.

64. *Massachusetts Spy*, December 31, 1772.

65. A draft in Adams's hand is part of the Samuel Adams Papers, NYPL.

66. Thomas, *Three Autobiographical Fragments*, 13.

67. Resolves of Ashford Town Committee, August 23, 1774, BCC Papers. The resolves appeared in the *Connecticut Gazette*, published by Timothy Green, on September 23, 1774.

68. See, for example, Peter Timothy to Samuel Adams, September 22, 1770, Samuel Adams Papers, NYPL; Timothy to S. Adams, June 9 and June 13, 1774, BCC Papers; S. Adams to Timothy, November 21, 1770, in *The Writings of Samuel Adams*, ed. Harry Alonzo Cushing (New York: Octagon Books, 1968 [1906]), 2:64–65; S. Adams to Timothy, July 27, 1774, Samuel Adams Papers, NYPL.

69. Samuel Adams to Arthur Lee, April 9, 1773, in Cushing, *The Writings of Samuel Adams*, 3:18–19.

70. On the company's corruption and the details of the plan, see Benjamin L. Carp, *Defiance of the Patriots: The Boston Tea Party & the Making of America* (New Haven, CT: Yale University Press, 2010), 7–24; Benjamin Woods Labaree, *The Boston Tea Party* (New York: Oxford University Press, 1964), 15–57; P. D. G. Thomas, *Tea Party to Independence: The Third Phase of the American Revolution, 1773–1776* (Oxford: Clarendon Press, 1991), 1–13.

71. Carp, *Defiance of the Patriots*, 73–77.

72. *Pennsylvania Packet*, October 28, 1771.

73. *Boston Gazette*, October 18, 1773; *Essex Gazette*, October 19, 1773; *Massachusetts Spy*, October 21, 1773; *Connecticut Gazette*, October 22, 1773; *Newport Mercury*, October 25, 1773; *Connecticut Courant*, October 26, 1773; *Pennsylvania Chronicle*, November 1, 1773.

74. *Pennsylvania Chronicle*, October 25, 1773; *Maryland Journal*, October 30, 1773; *Boston Evening Post*, November 8, 1773; *New-York Journal*, November 11, 1773.

75. *Rivington's New-York Gazetteer*, November 11, 1773; *Pennsylvania Chronicle*, November 15, 1773; *Pennsylvania Packet*, November 15, 1773; *Boston Post-Boy*, November 15, 1773; *Pennsylvania Gazette*, November 17, 1773; *New-York Journal*, November 18, 1773; *Norwich Packet*, November 18, 1773; *Essex Gazette*, November 23, 1773; *New-Hampshire Gazette*, November 26, 1773. The *Connecticut Journal* of November 19, 1773, published a summary of the piece. William Kelly, a New York merchant, had taken an active role in the late 1750s in suppressing informants against the illicit trade he and other merchants conducted during the Seven Years' War. See Thomas M. Truxes, *Defying Empire: Trading with the Enemy in Colonial New York* (New Haven, CT: Yale University Press, 2008), 12–16, 109–10.

76. *Rivington's New York Gazetteer*, November 18, 1773; *Pennsylvania Packet*, November 22, 1773; *Pennsylvania Journal*, November 24, 1773; *Boston Post-Boy*, November 22, 1773; *Boston Evening-Post*, November 29, 1773; *Newport Mercury*, November 29, 1773; *Providence Gazette*, December 4, 1773.

77. *Boston Post-Boy*, October 25, 1773; reprinted, *Massachusetts Spy*, October 28, 1773; *Boston Evening-Post*, November 1, 1773; *Essex Gazette*, November 2, 1773; *Connecticut Gazette*, November 5, 1773; *Connecticut Journal*, November 5, 1773; *Pennsylvania Journal*, November 10, 1773; *Pennsylvania Chronicle*, November 15, 1773; *Virginia Gazette* (Purdie & Dixon), November 25, 1773.

78. *Pennsylvania Journal*, November 17, 1773; *Pennsylvania Chronicle*, November 22, 1773.

79. *Rivington's New-York Gazetteer*, October 21, 1773; *Pennsylvania Chronicle*, October 25, 1773; *Pennsylvania Journal*, October 27, 1773; *Boston Post-Boy*, October 25, 1773.

80. Archer, *As If an Enemy's Country*.

81. Carp, *Defiance of the Patriots*, 117–40; Labaree, *Boston Tea Party*, 126–45.

82. *Boston Gazette*, November 29, 1773, *Boston Evening-Post*, November 29, 1773; reprinted *Essex Gazette*, November 30, 1773; *Connecticut Journal*, December 3, 1773; *Providence Gazette*, December 4, 1773; *Pennsylvania Journal*, December 8, 1773; *Pennsylvania Gazette*, December 9, 1773.

83. *Pennsylvania Gazette*, December 9, 1773.

84. Charles E. Clark, *The Public Prints: The Newspaper in Anglo-American Culture, 1665–1740* (New York: Oxford University Press, 1994).

85. *Boston Gazette*, December 20, 1773.

86. *Boston Evening-Post*, December 20, 1773.

87. Revere was a frequent courier for the Boston Committee of Correspondence, warning the residents of Portsmouth, New Hampshire, of an attempt by British troops to take a powder store in 1774 and, of course, warning Massachusetts towns about a British excursion in April 1775. See Jayne E. Triber, *A True Republican: The Life of Paul Revere* (Amherst: University of Massachusetts Press, 1998), 89–105.

88. Richard A. Ryerson, *The Revolution Is Now Begun: The Radical Committees of Philadelphia, 1765–1776* (Philadelphia: University of Pennsylvania Press, 1978), 33–38.

89. Toby L. Ditz, "Secret Selves, Credible Personas: The Problematics of Trust and Public Display in the Writings of Eighteenth-Century Philadelphia Merchants," in *Possible Pasts: Becoming Colonial in Early America*, ed. Robert Blair St. George (Ithaca, NY: Cornell University Press, 2000), 233–41; Kathleen Wilson, *The Sense of the People: Politics, Culture, and Imperialism in England, 1715–1785* (New York: Cambridge University Press, 1995).

90. "To the Delaware Pilots," November 27, 1773 (Philadelphia: n.p., 1773). Versions of the three letters were published in various newspapers: *New-York Journal*, December 2, 1773; *Boston Post-Boy*, December 6, 1773; *Newport Mercury*, December 20 and 27, 1773; *Essex Gazette*, December 21, 1773; *Providence Gazette*, December 25, 1773; *New-Hampshire Gazette*, January 7, 1774. On river pilots in Philadelphia during the Revolutionary era, see Simon Finger, "'A Flag of Defyance at the Masthead': The Delaware River Pilots and the Sinews of Philadelphia's Atlantic World in the Eighteenth Century," *Early American Studies* 8, no. 2 (Spring 2010): 386–409.

91. "Christmas-Box for the CUSTOMERS of the PENNSYLVANIA JOURNAL" (Philadelphia: Thomas Bradford, 1773). Hall and Sellers published the news as a "POSTSCRIPT to the *Pennsylvania Gazette*," December 24, 1773. See also Ryerson, *The Revolution Is Now Begun*, 37. By the late eighteenth century, the term "Christmas box" had shifted in meaning from a donation box for the poor to a general term for any Christmas present. Stephen Nissenbaum, *The Battle for Christmas: A Social and Cultural History of Our Most Cherished Holiday* (New York: Alfred A. Knopf, 1996), 110–12.

92. *Massachusetts Spy*, December 16, 1773.

93. Boston committee to Thomas Mifflin & George Clymer (Philadelphia), Philip Livingston and Samuel Broom (New York), December 17, 1773, Minute Book VI, 468–69, BCC Papers. The letter was first published in *Rivington's New-York Gazetteer*, December 23, 1773; reprinted, *Pennsylvania Gazette*, December 24, 1773; *Pennsylvania Chronicle*, December 27, 1773; *Pennsylvania Journal*, December 29, 1773; *Maryland Journal*, December 30, 1773; *Boston News-Letter*, December 30, 1773; *Boston Evening Post*, January 1, 1774; *Essex Gazette*, January 4, 1774; *Virginia Gazette* (Purdie & Dixon), January 6, 1774; *Providence Gazette*, January 8, 1774. See also Mary Beth Norton, "The Seventh Tea Ship," *William and Mary Quarterly*, 3rd ser., 73, no. 4 (2016): 681–710.

94. *Rivington's New-York Gazetteer*, December 23, 1773; *New-York Journal*, December 23, 1773; *Connecticut Journal*, December 24, 1773; *Pennsylvania Gazette*, December 24, 1773; *New-York Gazette, and Weekly Mercury*, December 27, 1773; *Pennsylvania Chronicle*, December 27, 1773; *Pennsylvania Journal*, December 29, 1773; *Virginia Gazette* (Purdie & Dixon), January 13, 1774. More detailed minutes of the action on December 16 first appeared in the *Boston Evening-Post*, December 20, 1773, and were reprinted in *Boston News-Letter*, December 23, 1773; *Massachusetts Spy*, December 23, 1773; *Connecticut Gazette*, December 24, 1773; *Providence Gazette*, December 25, 1773; *Boston Gazette*, December 27, 1773; *Essex Gazette*, December 28, 1773; *Norwich Packet*, December 30, 1773.

95. Jay David Bolter and Richard Grusin, *Remediation: Understanding New Media* (Cambridge, MA: MIT Press, 1999); quote on 45. My thanks to David Waldstreicher for introducing me to their work.

96. *Pennsylvania Gazette*, December 29, 1773.

97. *Pennsylvania Gazette*, December 29, 1773.

98. *Pennsylvania Journal*, December 29, 1773; *Pennsylvania Chronicle*, January 3, 1774; *Pennsylvania Packet*, January 3, 1774; *Wöchentliche Pennsylvanische Staatsbote*, January 4, 1774; *New-York Journal*, January 6, 1774; *Connecticut Gazette*, January 14, 1774; *Connecticut Journal*, January 14, 1774; *Boston Evening Post*, January 17, 1774; *Newport Mercury*, January 17, 1774; *Massachusetts Spy*, January 20, 1774; *Virginia Gazette* (Purdie & Dixon), January 20, 1774; *Providence Gazette*, January 22, 1774.

99. On Rivington, see Leroy Hewlett, "James Rivington, Loyalist Printer, Publisher, Bookseller of the American Revolution, 1724–1802: A Biographical-Bibliographical Study" (DLS diss., University of Michigan, 1958); Hewlett, "James Rivington, Tory Printer," in *Books in America's Past: Essays Honoring Rudolph H. Gjelsness*, ed. David Kaser (Charlottesville: University Press of Virginia, 1966), 165–94.

100. *Rivington's New-York Gazetteer*, October 28, 1773; reprinted, *Pennsylvania Packet*, November 1, 1773.

Chapter 4 · The Collision of Business and Politics, 1774–1775

1. P. D. G. Thomas, *Tea Party to Independence: The Third Phase of the American Revolution, 1773–1776* (Oxford: Clarendon Press, 1991), 26–87; David Ammerman, *In the Common Cause: American Response to the Coercive Acts of 1774* (Charlottesville: University Press of Virginia, 1974), 5–17.

2. May 13, 1774, Minute Book IX, 757–58, BCC Papers.

3. New Haven Committee of Correspondence to BCC, May 25, 1774, BCC Papers; Baltimore Committee of Correspondence to Dr. Taylor, May 25, 1774, BCC Papers.

4. Newspapers using the royal coat of arms as a visual part of their nameplate include such ideologically diverse prints as the *Connecticut Courant, Boston News-Letter, Boston Post-Boy, New-Hampshire Gazette*, and *Pennsylvania Gazette*, among others.

5. The eight pieces represented, head to tail: New England, New York, New Jersey, Pennsylvania, Maryland, Virginia, North Carolina, and South Carolina. See David Copeland, "'Join or Die': America's Newspapers in the French and Indian War," *Journalism History* 24, no. 3 (1998): 112–21.

6. T. H. Breen, *American Insurgents, American Patriots: The Revolution of the People* (New York: Hill and Wang, 2010), 192–93.

7. Joshua Brackett, Paul Revere, Benjamin Edes, Joseph Ward, Thomas Crafts Jr., and Thomas Chase to John Lamb, March 1, 1775, in Isaac Q. Leake, *Memoir of the Life and Times of General John Lamb* (Albany, 1850; repr., New York: Da Capo Press, 1971), 99–100.

8. Lyman Horace Weeks, *A History of Paper-Manufacturing in the United States, 1690–1916* (New York: Lockwood Trade Journal Co., 1916), 41–56.

9. Isaac Sears to Peter Vandevoort, October 20, 1774, Isaac Sears miscellaneous file, NYPL.

10. Francis Hutcheson to Thomas Bradford, April 15, 1774, Bradford Family Papers [coll. 1676], HSP.

11. See, for instance, William Strahan to William Hall, June 11, 1776, Letters to David Hall [coll. Am.162], HSP.

12. Ebenezer Hazard to Thomas Bradford, July 27, 1774, Ferdinand J. Dreer Autograph Collection [coll. 175], HSP.

13. *Massachusetts Spy*, September 29, 1774. Hazard eventually deferred the project until after the war, at which point he was undertaking a considerably more expansive compilation project of all the laws of the new states. See Fred Shelley, "Ebenezer Hazard: America's First Historical Editor," *William and Mary Quarterly*, 3rd ser., 12, no. 1 (1955): 44–73.

14. On the origins of the magazine genre in Britain, see James Tierney, "Periodicals and the Trade, 1695–1780," in *The Cambridge History of the Book in Britain*, vol. 5, *1695–1830*, ed. Michael F. Suarez and Michael L. Turner (Cambridge: Cambridge University Press, 2009), 479–97; Kathryn Shevelow, *Women and Print Culture: The Construction of Femininity in the Early Periodical* (New York: Routledge, 1989).

15. *Royal American Magazine, or Universal Repository of Instruction and Amusement* (Boston: Isaiah Thomas, January–June 1774; Joseph Greenleaf, July 1774–March 1775).

16. On the publication of essays in the *Royal American Magazine*, see E. W. Pitcher, "Sources for Fiction in the Royal American Magazine (Boston, 1774–5)," *American Notes & Queries* 17, no. 1 (1978): 6–7.

17. "Plan of the ROYAL AMERICAN MAGAZINE," *Massachusetts Spy*, June 24, 1773; reprinted in *Essex Gazette*, June 29, 1773; *Boston Post-Boy*, July 19,1773; *Newport Mercury*, July 26, 1773, August 2, 1773, August 9, 1773; *Pennsylvania Chronicle*, July 26, 1773, August 9, 1773; *Boston News-Letter*, July 29, 1773; *Providence Gazette*, July 31, 1773; *New-London Gazette*, August 27, 1773, September 10, 1773; *Connecticut Journal*, September 3, 1773, October 8, 1773.

18. See, in addition to Thomas's *Massachusetts Spy*: *Essex Gazette*, June 29, 1773, October 12, 1773, October 26, 1773; *Boston Post-Boy*, July 5, 1773, July 19, 1773, September 6, 1773, September 13, 1773, September 20, 1773, October 25, 1773, November 15, 1773, November 22, 1773; *New-York Journal*, July 22, 1773, August 5, 1773, November 11, 1773; *Newport Mercury*, July 26, 1773, August 2, 1773, August 9, 1773, November 15, 1773, December 13, 1773, December 20, 1773; *Pennsylvania Chronicle*, July 26, 1773, August 9, 1773, November 1, 1773, November 22, 1773, November 29, 1773; *Boston News-Letter*, July 29, 1773; *Providence Gazette*, July 31, 1773, November 20, 1773;

Boston Gazette, August 2, 1773, September 20, 1773, December 13, 1773; *Connecticut Gazette*, August 27, 1773, September 10, 1773, October 22, 1773, November 12, 1773; *Connecticut Journal*, September 3, 1773, October 8, 1773, October 22, 1773; *Boston Evening-Post*, September 13, 1773, November 1, 1773; *Pennsylvania Gazette*, October 27, 1773, November 10, 1773; *Norwich Packet*, November 18, 1773, November 25, 1773, December 2, 1773, December 16, 1773; *New-Hampshire Gazette*, November 26, 1773; *Essex Journal*, December 4, 1773.

19. *Royal American Magazine*, February 1774, table of contents. The May 1774 issue added William Goddard to the list, with offices in both Baltimore and Philadelphia.

20. Thomas Hutchinson, *The history of the colony of Massachusetts-Bay: from the first settlement thereof in 1628, until its incorporation with the colony of Plimouth, province of Main, &c. by the charter of King William and Queen Mary, in 1691*, 3rd ed. (Boston: Isaiah Thomas, 1774).

21. *Royal American Magazine*, June 1774, 202.

22. *Royal American Magazine*, July 1774, 234.

23. These issues are explored fully in Joseph M. Adelman, "'A Constitutional Conveyance of Intelligence, Public and Private': The Post Office, the Business of Printing, and the American Revolution," *Enterprise & Society* 11, no. 4 (2010): 709–52.

24. On Goddard, see Konstantin Dierks, *In My Power: Letter Writing and Commnications in Early America* (Philadelphia: University of Pennsylvania Press, 2009), 189; Daniel J. Boorstin, *The Americans: The Colonial Experience* (New York: Random House, 1958), 338; Richard R. John, *Spreading the News: The American Postal System from Franklin to Morse* (Cambridge, MA: Harvard University Press, 1995), 292–93; Isaiah Thomas, *The History of Printing in America, with a Biography of Printers, and an Account of Newspapers*, 2 vols. (Worcester, MA: Isaiah Thomas, Jr., 1810), 1:427–29, 2:63–66, 134–40; Ward L. Miner, *William Goddard: Newspaperman* (Durham, NC: Duke University Press, 1962).

25. *Maryland Journal*, August 20, 1773; *Massachusetts Spy*, March 17, 1774; *Connecticut Journal* (New Haven), March 25, 1774.

26. *Maryland Journal*, August 28, 1773; November 20, 1773.

27. William Goddard to [?], Baltimore, December 16, 1773, Lamb Papers, N-YHS.

28. Arthur M. Schlesinger, *Prelude to Independence: The Newspaper War on Britain, 1764–1776* (New York: Alfred A. Knopf, 1957), 192.

29. *Maryland Journal*, February 17, 1774.

30. *Pennsylvania Chronicle*, February 8, 1774.

31. For other perspectives on the Goddard proposal, see Schlesinger, *Prelude to Independence*, 190–95; Richard D. Brown, *Revolutionary Politics in Massachusetts: The Boston Committee of Correspondence and the Towns, 1772–1774* (Cambridge, MA: Harvard University Press, 1970), 181–84; Jerrilyn Greene Marston, *King and Congress: The Transfer of Political Legitimacy, 1774–1776* (Princeton, NJ: Princeton University Press, 1987), 228–30; T. H. Breen, *American Insurgents, American Patriots* (New York: Hill and Wang, 2010), 105–10; William A. Smith, *The History of the Post Office in British North America, 1639–1870* (Cambridge: Cambridge University Press, 1920; New York: Octagon Books, 1973), 60–63; Wesley Everett Rich, *The History of the United States Post Office to the Year 1829*, Harvard Economic Studies, vol. 27 (Cambridge, MA: Harvard University Press, 1924), 44–66.

32. Brown, *Revolutionary Politics in Massachusetts*, chaps. 7 and 8.

33. April 21, 1774, Minute-Book IX, 752, BCC Papers; "Nathaniel Appleton," in Clifford K. Shipton, *Sibley's Harvard Graduates*, 18 vols. (Boston: Massachusetts Historical Society, 1873–1999), 12:355–59.

34. "*The* PLAN *for establishing a New* American POST-OFFICE." The broadside was reprinted in such papers as the *Boston Post-Boy*, April 25, 1774; *Connecticut Journal*, April 29, 1774; *Boston Gazette*, May 2, 1774; *Massachusetts Spy*, May 5, 1774; *Providence Gazette*, May 7, 1774;

Essex Gazette, May 10, 1774; *Newport Mercury*, May 16, 1774; *Connecticut Courant*, May 31, 1774; *Virginia Gazette* (Purdie & Dixon), June 2, 1774.

35. "Subscription Paper relative to Post Office March 1774; (Indorsed by W. Cooper)," BCC Papers.

36. Pauline Maier, *From Resistance to Revolution: Colonial Radicals and the Development of American Opposition to Britain, 1765–1776* (New York: W. W. Norton, 1972); Brown, *Revolutionary Politics*; Edmund S. Morgan and Helen M. Morgan, *The Stamp Act Crisis: Prologue to Revolution* (Chapel Hill: IEAHC, University of North Carolina Press, 1953).

37. Thomas Young to John Lamb, May 13, 1774, quoted in Leake, *Memoir*, 86.

38. Brown, *Revolutionary Politics in Massachusetts*, 184.

39. March 17 and March 22, 1774, Minute Book IX, 733–34, BCC Papers.

40. T. H. Breen, *The Marketplace of Revolution: How Consumer Politics Shaped American Independence* (New York: Oxford University Press, 2004), 254–67.

41. *Massachusetts Spy*, March 17, 1774; *Essex Gazette*, March 22, 1774; *Connecticut Journal*, March 25, 1774.

42. Samuel Cutts to the Boston Committee of Correspondence, April 11, 1774, BCC Papers.

43. Richard D. Brown, *Knowledge Is Power: The Diffusion of Information in Early America, 1700–1865* (New York: Oxford University Press, 1989), 110–31.

44. Toby Ditz, "Secret Selves, Credible Personas: The Problematics of Trust and Public Display in the Writing of Eighteenth-Century Philadelphia Merchants," in *Possible Pasts: Becoming Colonial in Early America*, ed. Robert Blair St. George (Ithaca, NY: Cornell University Press, 2000), 219–42; Smith, *History of the Post Office*, 37–44.

45. Hugh Finlay, *Journal kept by Hugh Finlay, surveyor of the post roads on the continent of North America: during his survey of the post offices between Falmouth and Casco Bay, in the province of Massachusetts, and Savannah in Georgia, begun the 13th Septr., 1773 and ended 26th June 1774* (Brooklyn: F. H. Norton, 1867), 1–16; *Pennsylvania Chronicle*, October 11, 1773; *Pennsylvania Packet*, October 11, 1773; *New-York Gazette*, October 11, 1773; *Boston Post-Boy*, October 11, 1773; *Pennsylvania Gazette*, October 13, 1773; *Boston News-Letter*, October 14, 1773; *Connecticut Journal*, October 15, 1773; *Boston Gazette*, October 18, 1773; *Newport Mercury*, October 18, 1773; *Essex Gazette*, October 19, 1773.

46. *Massachusetts Spy*, March 17, 1774; *Boston Evening-Post*, March 21, 1774; *Connecticut Courant*, March 22, 1774; *Essex Gazette*, March 23, 1774; *Norwich Packet*, March 24, 1774; *Connecticut Gazette*, March 25, 1774; *Connecticut Journal*, March 25, 1774; *New-Hampshire Gazette*, March 25, 1774; *Virginia Gazette* (Purdie & Dixon), April 14, 1774.

47. *Boston Evening-Post*, March 21, 1774; *New-Hampshire Gazette*, March 25, 1774; *Boston Post-Boy*, March 28, 1774; *Essex Gazette*, March 29, 1774; *Connecticut Gazette*, April 1, 1774.

48. *Boston Post-Boy*, March 21, 1774; *Connecticut Courant*, March 22, 1774; *Essex Gazette*, March 22, 1774; *New-Hampshire Gazette*, March 25, 1774; *Connecticut Gazette*, March 25, 1774; *Connecticut Journal*, March 25, 1774; *New-Hampshire Gazette*, March 25, 1774; *Virginia Gazette* (Purdie & Dixon), April 14, 1774.

49. Boston committee to Newport and Providence committees, March 29, 1774, Minute Book IX, 748–49, BCC Papers.

50. Boston committee to Newport and Providence, March 29, 1774, Minute Book IX, 749, BCC Papers.

51. Benjamin Franklin to Thomas Cushing, September 15, 1774, *PBF* 21:306; Franklin to Jane Mecom, September 26, 1774, *PBF* 21:317.

52. Boston committee to Marblehead, Newburyport, Portsmouth, March 24, 1774, Minute Book IX, 734–36, BCC Papers. One legal scholar has even suggested that the "Constitutional Post" 's insistence on the secrecy of the mails is the basis for the Supreme Court's extension of

Fourth Amendment protections from searches to the arena of communications privacy. See Anuj C. Desai, "Wiretapping before the Wires: The Post Office and the Birth of Communications Privacy," *Stanford Law Review* 60 (2007): 553–94.

53. *Boston Post-Boy*, May 2, 1774; *Connecticut Journal*, May 6, 1774; *New-York Journal*, May 19, 1774; *Virginia Gazette* (Purdie & Dixon), June 2, 1774.

54. *Maryland Journal*, June 4, 1774.

55. John Foxcroft to Anthony Todd, April 5, 1774, National Archives, UK, T1/409/36. My thanks to Molly Warsh for transcribing this letter.

56. "Minutes of proceedings of John Foxcroft and Hugh Finlay, Deputy Postmasters General, relative to regulating the postal service in North America," November 24, 1774, Miscellaneous Collections: U.S. States and Territories, NYPL; *Pennsylvania Gazette*, August 3, 1774, August 10, 1774.

57. Debate notes of John Adams, October 7, 1775, *JCC* 3: 488–89.

58. Smith, *History of the Post Office*, 65.

59. Maya Jasanoff, *Liberty's Exiles: American Loyalists in the Revolutionary World* (New York: Alfred A. Knopf, 2011); Jerry Bannister and Liam Riordan, eds., *The Loyal Atlantic: Remaking the British Atlantic in the Revolutionary Era* (Toronto: University of Toronto Press, 2012); Robert M. Calhoon, Timothy M. Barnes, and George A. Rawlyk, eds., *Loyalists and Community in North America*, Contributions in American History, no. 158 (Westport, CT: Greenwood Press, 1994); Ruma Chopra, *Unnatural Rebellion: Loyalists in New York City during the Revolution*, Jeffersonian America (Charlottesville: University of Virginia Press, 2011); Joseph S. Tiedemann, *Reluctant Revolutionaries: New York City and the Road to Independence, 1763–1776* (Ithaca, NY: Cornell University Press, 1997); Judith L. Van Buskirk, *Generous Enemies: Patriots and Loyalists in Revolutionary New York*, Early American Studies (Philadelphia: University of Pennsylvania Press, 2002).

60. Joseph M. Adelman, "Trans-Atlantic Migration and the Printing Trade in Revolutionary America," *Early American Studies, An Interdisciplinary Journal* 11, no. 3 (Fall 2013): 539–44.

61. James Rivington to William and Thomas Bradford, November 17, 1774, and January 28, 1775, James Rivington collection [MS 2958.8286], N-YHS. The publications were [Thomas Jefferson], *A Summary View of the Rights of British America* (Williamsburg: Clementina Rind, 1774); [John Dickinson], *An Essay on the Constitutional Power of Great-Britain over the Colonies in America* (Philadelphia: William and Thomas Bradford, 1774); *Journal of the Proceedings of the Congress, Held at Philadelphia, September 5, 1774* (Philadelphia: William and Thomas Bradford, 1774). According to Thomas Adams, there were four editions of Jefferson's *Summary View*, all of which were printed in Williamsburg. Adams, *American Independence: The Growth of an Idea* (Providence, RI: Brown University Press, 1965), 90–91.

62. James Rivington to W. and T. Bradford, January 28, 1775, James Rivington collection [MS 2958.8286], N-YHS.

63. "James Rivington, Bookseller, Printer, and Stationer, In New-York. Proposes to publish a Weekly News-Paper . . ." February 15, 1773, LCP.

64. *Rivington's New-York Gazetteer*, April 22, 1773.

65. *Rivington's New-York Gazetteer*, January 13, 1774; February 10, 1774.

66. On the political situation in New York, see Joseph S. Tiedemann, *Reluctant Revolutionaries: New York City and the Road to Independence, 1763–1776* (Ithaca, NY: Cornell University Press, 1997); Judith L. Van Buskirk, *Generous Enemies: Patriots and Loyalists in Revolutionary New York* (Philadelphia: University of Pennsylvania Press, 2002).

67. *Rivington's New-York Gazetteer*, June 10, 1773.

68. *Rivington's New-York Gazetteer*, November 11, 1773; *Pennsylvania Chronicle*, November 15, 1773.

69. "Freinds of America" to Stephen Ward and Stephen Hopkins, December 5, 1774 (post-

marked at Newport, January 9, 1775), Ward Family Papers, ser. 4 [MS 776], Rhode Island Historical Society.

70. There is an annotation in the letter that refers Hopkins and Ward to "an Advertisement in his last Paper relative to A Proposed Raffle for Books Than which, can Any thing be more daring or Insolent." After scouring *Rivington's New-York Gazetteer* for the surrounding dates in December and January, I have been unable to locate the advertisement in question.

71. Ralph J. Randolph, "The End of Impartiality: 'South-Carolina Gazette,' 1763–75," *Journalism Quarterly* 49, no. 4 (September 1972): 706–9.

72. John Tobler, *The Georgia and South-Carolina Almanack, for the Year of Our Lord 1775 . . .*, Early American Imprints, 1st Ser. 13686 (Charleston [S.C.]: Printed for the editor, Sold in Georgia, by James Johnston at the printing-office in Savannah, Sold in South-Carolina, by Robert Wells, at the old printing-house, Great Stationary and Book Store in Charleston, 1774), http://nrs.harvard.edu/urn-3:hul.ebookbatch.EAIFS_batch:aas03005108.

73. Christopher Gould, "Robert Wells, Colonial Charleston Printer," *South Carolina Historical Magazine* 79, no. 1 (1978): 23–49.

74. Louisa Susannah Wells, *The Journal of a Voyage from Charlestown, S.C., to London: Undertaken during the American Revolution by a Daughter of an Eminent American Loyalist [Louisa Susannah Wells] [Aikman] in the Year 1778, and Written from Memory Only in 1779*, Divine Jones Fund Series of Histories and Memoirs, no. 2 (New York: Printed for the New York Historical Society, 1906).

75. Thomas, *History of Printing in America*, 2:159.

76. Ammerman, *In the Common Cause*, 73–88.

77. Robert G. Parkinson, *The Common Cause: Creating Race and Nation in the American Revolution* (Chapel Hill: OIEAHC, University of North Carolina Press, 2016), 78–97; Brown, *Knowledge Is Power*, 247–53, quote at 249. See also Frank Luther Mott, "The Newspaper Coverage of Lexington and Concord," *New England Quarterly* 17, no. 4 (1944): 489–505.

78. Joseph Palmer, "To all friends of American liberty . . .," April 19, 1775 [coll. Am.606], HSP. On the letter and its distribution, see John H. Scheide, "The Lexington Alarm," *Proceedings of the American Antiquarian Society*, n.s., 50 (1940): 49–79. On the distribution of the news of Lexington and Concord from Boston generally, see Lester J. Cappon, ed., *Atlas of Early American History: The Revolutionary Era, 1760–1790* (Princeton, NJ: Published for the Newberry Library and IEAHC by Princeton University Press, 1976), 42.

79. Allan R. Pred identified a standard time lag of about two weeks for news to travel from Boston to Philadelphia in a series of time-lag studies on the early republic. "Urban Systems Development and the Long-Distance Flow of Information through Preelectronic U.S. Newspapers," *Economic Geography* 47, no. 4 (1971): 510. On the arrival of the news in Philadelphia, see Richard A. Ryerson, *The Revolution Is Now Begun: The Radical Committees of Philadelphia, 1765–1776* (Philadelphia: University of Pennsylvania Press, 1978), 117–18.

80. As it was reproduced, the substantive details of the letter remained the same as it was in manuscript and first printings. However, some facts, particularly the spelling of names, fell victim to the game of "telephone" that occurred. Most notably, the name of the first rider from Massachusetts, Israel Bissell, was reproduced in various forms. In the New York papers (the *New-York Gazette* and *Rivington's New-York Gazetteer*) he was "Israel Bessel." By the time news reached Philadelphia, the name had been transposed in manuscript and in print to "Trail Bissel." See *Pennsylvania Packet*, April 24, 1775; *Pennsylvania Evening-Post*, April 25, 1775; *Pennsylvania Gazette*, April 26, 1775; *Pennsylvania Mercury*, April 28, 1775. All of these seem to stem from problems in interpreting handwriting. However, it is unclear how his name became "Tryal Russell" between Philadelphia and Baltimore. See "Baltimore: April 26. We have just received the following important intelligence" (Baltimore: n.p., 1775).

81. "PHILADELPHIA, *April* 24, 1775. *An Express arrived at Five o'Clock this Evening, by*

which we have the following Advices" (Philadelphia: William and Thomas Bradford, 1775), HSP; "Philadelphia. April 25th, 1775. An express arrived at five o'clock this evening, by which we have the following advices: Watertown, Wednesday morning, near ten of the clock. To all friends of American liberty . . ." (Lancaster, PA: Francis Bailey, 1775).

82. *Norwich Packet*, April 20, 1775.

83. *New-York Gazette, and Weekly Mercury*, April 24, 1775; *Rivington's New-York Gazetteer*, April 27, 1775. It is likely that Rivington, who published the letter three days after Gaine, simply reproduced his words.

84. *The Literary Diary of Ezra Stiles, D.D. LL.D.*, ed. Franklin Bowditch Dexter, 3 vols. (New York: Charles Scribner's Sons, 1901), 1:558. The Halls published the *New England Chronicle* until September 1776.

85. "At a Meeting of the Convention of Committees of the County of Worcester on Wednesday the 31. Day of May AD. 1775 at the Court House in Worcester," United States Revolution Collection, AAS.

86. John Hancock to Isaiah Thomas, April 4, 1775, Isaiah Thomas Papers, AAS. The ad appeared on April 6, 1775 both in the *Massachusetts Spy* and in the *Boston News-Letter* of Margaret Draper.

87. Isaiah Thomas to Daniel Hopkins, October 2, 1775, Isaiah Thomas Papers, AAS.

88. *Diary of Ezra Stiles*, May 6, 1775, 1:546–57.

89. *Diary of Ezra Stiles*, September 16, 1776, 2:54.

90. Journal of Nicholas Cresswell, October 9, 1776, in *A Man Apart: The Journal of Nicholas Cresswell, 1774–1781*, ed. Harold B. Gill Jr. and George M. Curtis III (Lanham, MD: Lexington Books, 2009), 127. See also November 24, 1776, in *A Man Apart*, 131.

Chapter 5 · Patriots, Loyalists, and the Perils of Wartime Printing

1. James A. Henretta, "The War for Independence and American Economic Development," in *The Economy of Early America: The Revolutionary Period, 1763–1790*, ed. Ronald Hoffman, Perspectives on the American Revolution (Charlottesville: Published for the United States Capitol Historical Society by the University Press of Virginia, 1988), 45–87; John McCusker and Russell Menard, *The Economy of British America, 1607–1789* (Chapel Hill: IEAHC, University of North Carolina Press, 1985), 351–77.

2. A total of 184 printers were active during the years 1775 to 1783, which means that 78 have indeterminate political leanings. It is certainly likely, given estimates of the ratio of Patriots, Loyalists, and neutrals, that many of these other printers either had no strong inclinations or ran offices of such insignificance that no one was curious. On the ratio of Patriots, Loyalists, and neutrals, see Robert M. Calhoon, "Loyalism and Neutrality," in *The Blackwell Encyclopedia of the American Revolution*, ed. Jack P. Greene and J. R. Pole (Cambridge, MA: Blackwell, 1991), 247–59. On printers during the American Revolution, see Bernard Bailyn and John B. Hench, eds., *The Press and the American Revolution* (Worcester, MA: American Antiquarian Society, 1980), 229–72. On Loyalists, see George Edward Cullen Jr., "Talking to a Whirlwind: The Loyalist Printers in America, 1763–1783" (PhD diss., West Virginia University, 1979); Timothy M. Barnes, "Loyalist Newspapers of the American Revolution, 1763–1783: A Bibliography," *Proceedings of the American Antiquarian Society* 83 (1973): 217–40.

3. John William Wallace, *An Old Philadelphian, Colonel William Bradford, the Patriot Printer of 1776. Sketches of His Life* (Philadelphia: Sherman & Co., Printers, 1884), 120–227.

4. G. Thomas Tanselle, "Some Statistics on American Printing, 1764–1783," in Bailyn and Hench, *The Press and the American Revolution*, 315–63; E. James Ferguson, *The Power of the Purse: A History of American Public Finance, 1776–1790* (Chapel Hill: IEAHC, University of North Carolina Press, 1961), 25–47.

5. On Brackenridge, see Daniel Marder, *Hugh Henry Brackenridge* (New York: Twayne

Publishers, 1967); Joseph J. Ellis, *After the Revolution: Profiles of Early American Culture* (New York: W. W. Norton, 1979), 73–110. On his editing of the *United States Magazine*, see Claude Milton Newlin, *The Life and Writings of Hugh Henry Brackenridge* (Princeton, NJ: Princeton University Press, 1932), 44–57.

6. *The United States Magazine; A Repository of History, Politics and Literature* (Philadelphia: printed by Francis Bailey), December 1779, 483.

7. William Strahan to William Hall, July 5, 1775, Letters to David Hall (Am.162), HSP.

8. Peter Timothy to Henry Laurens, February 14, 1779, Peter Timothy Miscellaneous File, NYPL.

9. Samuel Loudon to Matthew Visscher, August 20, 1777, September 4, 1777, Samuel Loudon Papers, N-YHS.

10. See, for example, discussions on establishing communication with New York, July 5, 1776, *JCC* 5:522; with the southern colonies, August 29, 1776, 5:717–18; with Fort Ticonderoga, September 3, 1776, 5:732, November 5, 1776, 6:926–27; with the army, October 31, 1776, 6:916; Konstantin Dierks, *In My Power: Letter Writing and Communications in Early America* (Philadelphia: University of Pennsylvania Press, 2009), 206–14.

11. George Washington to John Hancock, September 4, 1776, Papers of the Continental Congress (National Archives Microfilm Publication M247, roll 186), item 169, 2:202–5.

12. Ezra Stiles, September 16, 1776, in *The Literary Diary of Ezra Stiles, D.D. LL.D.*, ed. Franklin Bowditch Dexter (New York: Charles Scribner's Sons, 1901), 2:54.

13. Declaration of Independence (Baltimore: Mary Katherine Goddard, 1777), Early American Imprints, ser. 1, no. 15650.

14. Rollo G. Silver, "Aprons Instead of Uniforms: The Practice of Printing, 1776–1787," *Proceedings of the American Antiquarian Society* 87, no. 1 (1977): 121.

15. Silver, "Aprons Instead of Uniforms," 142–43. Kollock forewent the custom of publishing an address in the first number of the *New-Jersey Journal* (February 16, 1779), which makes it difficult to evaluate his publicly stated motives.

16. Trenton C. Jones, "Displaying the Ensigns of Harmony: The French Army in Newport, Rhode Island, 1780–1781," *New England Quarterly* 85, no. 3 (2012): 430–67; Eugena Poulin and Claire Quintal, eds., *La Gazette Françoise: Revolutionary America's French Newspaper* (Newport, RI: Salve Regina University, in association with University Press of New England, 2007). According to Poulin and Quintal, most of the material for the French newspaper came from the *Newport Mercury*.

17. Steven Carl Smith, *An Empire of Print: The New York Publishing Trade in the Early American Republic*, Penn State Series in the History of the Book (University Park: Pennsylvania State University Press, 2017), 7–43; Silver, "Aprons Instead of Uniforms," 131–35.

18. Smith, *An Empire of Print*, 141.

19. Smith, *An Empire of Print*, 144.

20. Dexter, *The Literary Diary of Ezra Stiles*, 2:549.

21. Maurice R. Cullen Jr., "Benjamin Edes: Scourge of Tories," *Journalism Quarterly* 51 (1974): 213–18; Rollo G. Silver, "Benjamin Edes: Trumpeter of Sedition," *Papers of the Bibliographical Society of America* 47 (1953): 248–68.

22. Isaiah Thomas, *Three Autobiographical Fragments; Now First Published upon the 150th Anniversary of the Founding of the American Antiquarian Society, October 24, 1812* (Worcester, MA: American Antiquarian Society, 1962), 14.

23. Warrant for arrest of Isaiah Thomas, Suffolk County, Massachusetts, April 13, 1775, Tileston & Hollingsworth Co. Papers, AAS.

24. John Boyle, *A Journal of Occurrences in Boston, 1759–1778*, 1:147, Ms Am 1926, Houghton Library, Harvard University, Cambridge, MA.

25. Goddard's attempt also met resistance from Pennsylvania delegate Joseph Galloway.

October 5, 1774, *JCC* 1:55; Diary of Silas Deane, October 5, 1774, *Letters of Delegates to Congress, 1774–1789*, 25 vols., ed. Paul H. Smith (Washington, DC: Library of Congress, 1976–98), 1:143.

26. *New-Hampshire Gazette*, May 19, 1775, June 2, 1775; *Connecticut Gazette*, June 9, 1775; *New-York Journal*, June 1, 1775 (reprinted weekly until July 27, 1775); William Bradford, "An Account for Postages, May 11, 1775–July 27, 1775," Bradford Family Papers, Series 2, HSP.

27. *New-Hampshire Gazette*, June 2, 1775.

28. May 29, 1775, *JCC* 2:71; July 26, 1775, *JCC* 2:208–9.

29. Samuel Ward to Franklin, August 12, 1775, *PBF* 22:167–69; Note, *PBF* 22:183.

30. *Virginia Gazette* (Purdie), October 13, 1775.

31. Christopher J. Young, "Mary K. Goddard: A Classical Republican in a Revolutionary Age," *Maryland Historical Magazine* 96 (2001): 4–27; Richard R. John and Christopher J. Young, "Rites of Passage: Postal Petitioning as a Tool of Governance in the Age of Federalism," in *The House and the Senate in the 1790s: Petitioning, Lobbying, and Institutional Development*, ed. Kenneth R. Bowling and Donald R. Kennon (Athens: Ohio University Press for the Unites States Capitol Historical Society, 2002), 109–14.

32. *New-York Journal*, June 1, 1775 (reprinted weekly until July 27, 1775).

33. *Massachusetts Spy*, May 10, 1775; *New-Hampshire Gazette*, May 19, 1775.

34. Thomas Bradford to John Holt, "draft," July 27, 1775, Society Collection, HSP.

35. Ward L. Miner, *William Goddard: Newspaperman* (Durham, NC: Duke University Press, 1962), 135.

36. Franklin to Silas Deane, August 27, 1775, *PBF* 22:183–85.

37. May 12, 1777, *JCC* 7:347.

38. Jan. 11, 1777, *JCC* 7:29–30.

39. John A. Nagy, *Invisible Ink: Spycraft of the American Revolution* (Yardley, PA: Westholme, 2010), 21–26.

40. John Adams to Abigail Adams, July 24, 1775; John Adams to James Warren, July 24, 1775; The *Adams Papers Digital Edition*, ed. C. James Taylor (Charlottesville: University of Virginia Press, Rotunda, 2008). For more, see Russ Castronovo, *Propaganda 1776: Secrets, Leaks, and Revolutionary Communications in Early America*, Oxford Studies in American Literary History (New York: Oxford University Press, 2014), 42–44; *Massachusetts Gazette and Boston News-Letter*, August 17, 1775.

41. *Norwich Packet*, January 15, 1776; Dexter, *The Literary Diary of Ezra Stiles*, 1:652, January 6, 1776; Harold B. Gill Jr. and George M. Curtis III, eds., *A Man Apart: The Journal of Nicholas Cresswell, 1774–1781* (Lanham, MD: Lexington Books, 2009), July 9, 1776, 111–12.

42. "*The* PLAN *for establishing a New* American POST-OFFICE" (Boston: n.p., 1774). The broadside was reprinted in such papers as the *Boston Post-Boy*, April 25, 1774; *Connecticut Journal*, April 29, 1774; *Boston Gazette*, May 2, 1774; *Massachusetts Spy*, May 5, 1774; *Providence Gazette*, May 7, 1774; *Essex Gazette*, May 10, 1774; *Newport Mercury*, May 16, 1774; *Connecticut Courant*, May 31, 1774; *Virginia Gazette* (Purdie & Dixon), June 2, 1774.

43. Ian K. Steele, *The English Atlantic, 1675–1740: An Exploration of Communication and Community* (New York: Oxford University Press, 1986), 64.

44. Benjamin Hawkins and Hugh Williamson to Alexander Martin, January 28, 1783, in Smith, *Letters of Delegates to Congress*, 19:633–34.

45. Hawkins and Williamson to Martin, January 28, 1783, in Smith, *Letters of Delegates to Congress*, 19:633–34.

46. Charles Royster, *A Revolutionary People at War: The Continental Army & American Character, 1775–1783* (Chapel Hill: IEAHC, University of North Carolina Press, 1979), 25–53.

47. Pauline Maier, *American Scripture: Making the Declaration of Independence* (New York: Knopf, 1997), 21.

48. "Account of the Manufacture of Salt-Petre," *Pennsylvania Magazine* (June 1775), 266–68.

49. "Observations on the Military Character of Ants," *Pennsylvania Magazine* (July 1775), 300. See also Edward Larkin, *Thomas Paine and the Literature of Revolution* (Cambridge: Cambridge University Press, 2005), 22–24.

50. Larkin, *Thomas Paine and the Literature of Revolution*, 55; Richard D. Brown, *The Strength of a People: The Idea of an Informed Citizenry in America, 1650–1870* (Chapel Hill: University of North Carolina Press, 1996), 64–66. For a fuller explanation of Paine's argument, see Eric Foner, *Tom Paine and Revolutionary America* (New York: Oxford University Press, 1986), 71–106. For more on Paine, see Trish Loughran, "Disseminating *Common Sense*: Thomas Paine and the Problem of the Early National Bestseller," *American Literature* 78, no. 1 (March 2006): 1–28; Loughran, *The Republic in Print: Print Culture in the Age of U.S. Nation Building, 1770–1870* (New York: Columbia University Press, 2007), chap. 2; Michael Everton, "'The Would-Be-Author and the Real Bookseller': Thomas Paine and Eighteenth-Century Printing Ethics," *Early American Literature* 40, no. 1 (2005): 79–110; Richard A. Ryerson, *The Revolution Is Now Begun: The Radical Committees of Philadelphia, 1765–1776* (Philadelphia: University of Pennsylvania Press, 1978), 152–55. On Paine's career and ideology, see Seth Cotlar, *Tom Paine's America: The Rise and Fall of Transatlantic Radicalism in the Early Republic* (Charlottesville: University of Virginia Press, 2011); Foner, *Tom Paine and Revolutionary America*; Nicole Eustace, *Passion Is the Gale: Emotion, Power, and the Coming of the American Revolution* (Chapel Hill: OIEAHC, University of North Carolina Press, 2008), 439–79; Jack Fruchtman Jr., *The Political Philosophy of Thomas Paine* (Baltimore: Johns Hopkins University Press, 2009); Fruchtman, *Thomas Paine: Apostle of Freedom* (New York: Four Walls Eight Windows, 1994); Harvey J. Kaye, *Thomas Paine and the Promise of America* (New York: Hill and Wang, 2005); John Keane, *Tom Paine: A Political Life* (London: Bloomsbury, 1995); David Freeman Hawke, *Paine* (New York: Harper and Row, 1974). On Bell, see Sarah Knott, *Sensibility and the American Revolution* (Chapel Hill: OIEAHC, University of North Carolina Press, 2008); James N. Green, "English Books and Printing in the Age of Franklin," in *CBAW*, 283–89.

51. Thomas Paine, *Common Sense, addressed to the inhabitants of America, on the following interesting subjects*, new ed. (Philadelphia: William and Thomas Bradford, 1776), 11.

52. Paine, *Common Sense*, 37.

53. Loughran, *The Republic in Print*, 33–103.

54. Loughran, *The Republic in Print*, 38–42.

55. Thomas R. Adams cataloged twenty-five American editions printed in thirteen towns. Editions were also published in London, Edinburgh, Newcastle-upon-Tyne, and Rotterdam. Adams, *American Independence: The Growth of an Idea; A Bibliographical Study of the American Political Pamphlets Printed between 1764 and 1776 Dealing with the Dispute between Great Britain and Her Colonies*, Brown University Bicentennial Publications: Studies in the Fields of General Scholarship (Providence, RI: Brown University Press, 1965), xi, 164–72; Loughran, *The Republic in Print*, 47.

56. For example, see *Norwich Packet*, March 4, 1776, March 11, 1776, March 18, 1776, March 25, 1776, April 1, 1776, April 8, 1776, April 15, 1776, April 22, 1776; *Connecticut Courant*, February 26, 1776, March 4, 1776; *Virginia Gazette* (Purdie), February 2, 1776; *Virginia Gazette* (Pinkney), February 3, 1776.

57. Horatio Gates to Charles Lee, January 22, 1776, Sol Feinstone Collection, DLAR (on deposit at APS). Gates was not alone in his estimation of Franklin's involvement.

58. William Whitcroft to Thomas Bradford, February 19, 1776, Bradford Family Papers, HSP; John Carter to Joseph Trumbull, March 6, 1776, Book Trades Collection, AAS. On the publication history of the pamphlet, see Loughran, "Disseminating *Common Sense*"; Richard Gimbel, *Thomas Paine: A Bibliographical Check List of* Common Sense *with an Account of Its Publication* (New Haven, CT: Yale University Press, 1956).

59. *New-England Chronicle*, April 4, 1776; reprinted in *Essex Journal*, April 5, 1776; *Con-*

necticut Journal, April 10, 1776; *New-York Journal*, April 11, 1776; *Pennsylvania Packet*, April 15, 1776; *Newport Mercury*, April 22, 1776; *Virginia Gazette* (Purdie), April 26, 1776; *Virginia Gazette* (Dixon & Hunter), April 27, 1776.

60. Extracts of the pamphlet appeared, for example, in the *London Evening Post*, May 30, June 1, 1776; *General Evening Post*, June 4, June 8, 1776, reprinted in *Middlesex Journal and Evening Advertiser*, June 4, 1776, and *Craftsman or Say's Weekly Journal*, June 8, 1776. Excerpts from "Additions to Common Sense" appeared in *Gazetteer and New Daily Advertiser*, June 22, June 28, June 29, 1776; *General Evening Post*, July 6, 1776. Advertisements for the *Common Sense* and the Additions appeared in the *Gazetteer and New Daily Advertiser*, May 31, June 6, June 8, June 11, 1776; *London Evening Post*, June 4, June 20, June 22, July 2, 1776; *Morning Post and Daily Advertiser*, June 14, June 28, July 5, July 22, 1776; *St. James's Chronicle and the British Evening Post*, August 6, 1776.

61. Gill and Curtis, *A Man Apart:*, January 19, 1776, 104.

62. "Candidus," *Plain Truth; Addressed to the Inhabitants of America, Containing, Some Remarks on a Late Pamphlet entitled Common Sense* (Philadelphia: Robert Bell, 1776).

63. Larkin, *Thomas Paine and the Literature of Revolution*, 54–55.

64. Paine outed himself as the author of *Common Sense* in a series of letters signed as "Forester," written in response to "Cato" (William Smith). See *Pennsylvania Evening Post*, April 30, 1776. Philip Gould, *Writing the Rebellion: Loyalists and the Literature of Politics in British America*, Oxford Studies in American Literary History (New York: Oxford University Press, 2013), 114–43.

65. Adams, *American Independence*, 152–53.

66. Pauline Maier, *American Scripture: Making the Declaration of Independence* (New York: Alfred A. Knopf, 1997), 28–34; Robert G. Parkinson, *The Common Cause: Creating Race and Nation in the American Revolution* (Chapel Hill: OIEAHC, University of North Carolina Press, 2016), 189–95.

67. Danielle S. Allen, *Our Declaration: A Reading of the Declaration of Independence in Defense of Equality* (New York: Liveright Publishing, 2014); Maier, *American Scripture*; Steven C. A. Pincus, *The Heart of the Declaration: The Founders' Case for an Activist Government*, Lewis Walpole Series in Eighteenth-Century Culture and History (New Haven, CT: Yale University Press, 2016); David Armitage, *The Declaration of Independence: A Global History* (Cambridge, MA: Harvard University Press, 2007). I would also like to thank Emily Sneff and Danielle Allen of the Declaration Resources Project at Harvard University for their assistance with references and information about the publication history of the Declaration.

68. Parkinson, *The Common Cause*, 229–61.

69. Liz Covart, *A Declaration in Draft*, Ben Franklin's World, accessed July 4, 2017, https://www.benfranklinsworld.com/episode-141-declaration-draft/; Allen, *Our Declaration*.

70. It is unclear how many copies Dunlap printed, though most historians estimate the print run at between 150 and 200. See Emily Sneff, "The Substance and Style of the 1776 Newspaper and Broadside Editions of the Declaration of Independence" (unpublished manuscript, April 6, 2018), available at http://declaration.fas.harvard.edu/resources/1776-editions.

71. "When and How Did the Colonies Find Out about the Declaration?," Declaration Resources Project, https://declaration.fas.harvard.edu/resources/when-how.

72. Armitage, *The Declaration of Independence*; Eliga H. Gould, *Among the Powers of the Earth: The American Revolution and the Making of a New World Empire* (Cambridge, MA: Harvard University Press, 2014).

73. Thomas Starr, "Separated at Birth: Text and Context of the Declaration of Independence," *Proceedings of the American Antiquarian Society* 110, no. 1 (2000): 163.

74. On the effect of military campaigns on communications, see Parkinson, *The Common Cause*.

75. Ferguson, *The Power of the Purse*, 3–24.

76. Carol Sue Humphrey, *"This Popular Engine": New England Newspapers during the American Revolution* (Newark: University of Delaware Press, 1992), 23–43; Lyman Horace Weeks, *A History of Paper-Manufacturing in the United States, 1690–1916* (New York: Lockwood Trade Journal Co., 1916), 46–55; Eugenie Andruss Leonard, "Paper as a Critical Commodity during the American Revolution," *Pennsylania Magazine of History and Biography* 74, no. 4 (1950): 488–99.

77. Samuel Loudon to Isaac Beers, November 3, 1777, Samuel Loudon Papers, N-YHS.

78. References in this paragraph are based on observation of the *Maryland Journal* held by the American Antiquarian Society in 2015. For more information on AAS holdings, see http://clarence.mwa.org/Clarence/full-holding.php?bib_id=136.

79. The line was "Omne tulit punctum qui miscuit utile dulci, lectorem delectando pariterque monendo," from Horace's *Ars Poetica*, which translates to "He who joins the instructive with the agreeable, carries off every vote, by delighting and at the same time admonishing the reader." See https://www.poetryfoundation.org/articles/69381/ars-poetica.

80. See, for example, *Maryland Journal*, October 2, October 16, 1776.

81. *Maryland Journal*, Apr. 8, 1777. On the possible identity of the paper mill, see Bidwell, *American Paper Mills*, 230.

82. *Maryland Journal*, May 5, 1778.

83. Silver, "Aprons Instead of Uniforms," 112–31.

84. Ruma Chopra, "Printer Hugh Gaine Crosses and Re-crosses the Hudson," *New York History* 90, no. 4 (2009): 271–85.

85. Janice Potter and Robert M. Calhoon, "The Character and Coherence of the Loyalist Press," in Bailyn and Hench, *The Press and the American Revolution*, 229–72.

86. Ambrose Serle to the Earl of Dartmouth, November 26, 1776, in Paul Leicester Ford, ed., *The Journals of Hugh Gaine, Printer* (New York: Dodd, Mead, 1902), 1:57–58; Parkinson, *Common Cause*, 316–22. On Serle's editing of the Gaine *New-York Gazette*, see Chopra, "Printer Hugh Gaine Crosses and Re-crosses the Hudson," 283.

87. Dierks, *In My Power*, 209.

88. On Collins, see Richard F. Hixson, *Isaac Collins: A Quaker Printer in 18th Century America* (New Brunswick, NJ: Rutgers University Press, 1968); *Memoir of the Late Isaac Collins, of Burlington, New Jersey* (Philadelphia: printed by Joseph Rakestraw, 1848).

89. On violence against printers generally, see John C. Nerone, *Violence against the Press: Policing the Public Sphere in U.S. History* (New York: Oxford University Press, 1994), 18–52.

90. Parkinson, *The Common Cause*, 144–45.

91. *Virginia Gazette* (Dixon & Hunter), October 7, 1775.

92. *Maryland Journal*, February 25, 1777. For a general narrative of the events, see Miner, *William Goddard*, 150–62.

93. "Memorial of William Goddard," March 6, 1777, Goddard Family Papers [MSS 442], Rhode Island Historical Society, Providence.

94. "Deposition of Andrew Wilson," April 2, 1777, Goddard Family Papers [MSS 442], Rhode Island Historical Society.

95. Isaac Sears to Roger Sherman, Eliphalet Dyer, and Silas Deane, November 28, 1775, Sol Feinstone Collection, DLAR. A letter in the *Constitutional Gazette* (New York) of November 29, 1775, identified three Loyalists as particular targets: "Parson Seabury, Judge Fowler, and Lord Underhill." See also Holger Hoock, *Scars of Independence: America's Violent Birth* (New York: Crown, 2017), 38–39; Parkinson, *The Common Cause*, 176–79; Dwight L. Teeter, "'King' Sears, the Mob and Freedom of the Press in New York, 1765–1776," *Journalism Quarterly* 41, no. 4 (1964): 539–44; Leroy Hewlett, "James Rivington, Tory Printer," in *Books in America's Past: Essays Honoring Rudolph H. Gjelsness*, ed. David Kaser (Charlottesville: University Press of Virginia, 1966), 172–74.

96. *Constitutional Gazette* (New York), November 25, 1775; reprinted in *Pennsylvania Gazette*, November 29, 1775; *Connecticut Journal*, November 29, 1775; *New-England Chronicle*, December 7, 1775; *Essex Journal*, December 8, 1775; *Boston Gazette*, December 11, 1775. Other versions of the story appeared in *Connecticut Journal*, November 29, 1775; *New-England Chronicle*, November 30, 1775; *Virginia Gazette* (Purdie), December 8, 1775.

97. *Virginia Gazette* (Purdie), December 8, 1775.

98. Sears to Sherman, Dyer, and Deane, November 28, 1775, Feinstone Collection, DLAR.

99. *New-York Gazette, and Weekly Mercury*, January 15, 1776; reprinted in *Connecticut Journal*, January 18, 1776; *Pennsylvania Evening-Post*, January 18, 1776; *Pennsylvania Ledger*, January 20, 1776; *Pennsylvania Packet*, January 22, 1776.

100. Catherine Snell Crary, "The Tory and the Spy: The Double Life of James Rivington," *William and Mary Quarterly*, 3rd ser., 16, no. 1 (1959): 61–72; Nagy, *Invisible Ink*, 245–46; Kenneth Daigler, *Spies, Patriots, and Traitors: American Intelligence in the Revolutionary War* (Washington, DC: Georgetown University Press, 2014), 186–87.

101. *United States Magazine*, January 1779, 34–40. For the attribution to Witherspoon, see Claude Milton Newlin, *The Life and Writings of Hugh Henry Brackenridge* (Princeton, NJ: Princeton University Press, 1932), 48.

102. *United States Magazine*, January 1779, 36.

103. On the Robertsons, see Marion Robertson, "The Loyalist Printers: James and Alexander Robertson," *Nova Scotia Historical Review* 3, no. 1 (1983): 83–93.

104. Claim of James and Alexander Robertson, March 25, 1784, American Loyalist Claims, National Archives, UK, AO 12/19/280. It appears that the Robertson brothers received approximately £250 in compensation. AO 12/109/256–57.

105. Claim of James and Alexander Robertson, March 25, 1784, 282.

106. Frank Cundall, *A History of Printing in Jamaica from 1717 to 1834* (Kingston: Institute of Jamaica, 1935), 32–33.

107. Memorial of Robert Luist Fowle, Papers of the Loyalist Claims Commission, AO 13/52/234, viewed at DLAR.

108. On the Loyalist Claims Commission and the Loyalist diaspora, see Maya Jasanoff, *Liberty's Exiles: American Loyalists in the Revolutionary World* (New York: Alfred A. Knopf, 2011); Alan Taylor, *American Revolutions: A Continental History* (New York: W. W. Norton., 2016), 323–27.

109. Taylor, *American Revolutions*, 284–85.

110. Claim of James Humphreys, February 22, 1784, American Loyalist Claims, National Archives, UK, AO 12/38/101–2, viewed at DLAR.

111. Sidney E. Berger, "Innovation and Diversity among the Green Family of Printers," *Printing History* 12, no. 1 (1990): 2–20; Leona M. Hudak, *Early American Women Printers and Publishers, 1639–1820* (Metuchen, NJ: Scarecrow Press, 1978), 397–423; Richard L. Demeter, *Primer, Presses, and Composing Sticks: Women Printers of the Colonial Period* (Hicksville, NY: Exposition Press, 1979), 119–35.

112. James P. O'Donnell, "Richard Draper," in *Boston Printers, Publishers, and Booksellers, 1640–1800*, ed. Benjamin Franklin V (Boston: G. K. Hall, 1980), 106–12.

113. Isaiah Thomas, *The History of Printing in America, with a Biography of Printers, and an Account of Newspapers* (Worcester, MA: Isaiah Thomas, Jr., 1810), 2:207–8, 245.

114. Boyle, *Journal of Occurrences*, April 19, 1774 [MS Am 1926], Houghton Library, Harvard University.

115. Boyle, *Journal of Occurrences*, August 4, 1774 [MS Am 1926], Houghton Library, Harvard University.

116. Howe later purchased some of the printing equipment Draper had taken with her to

Nova Scotia. Marie Tremaine, *A Bibliography of Canadian Imprints, 1751–1800* (Toronto: University of Toronto Press, 1952), 662.

117. Petition of Margaret Draper, Massachusetts, American Loyalist Claims, National Archives, UK, AO 13/44/341, viewed at Ancestry.com.

118. Petition of Margaret Draper, Massachusetts, American Loyalist Claims, National Archives, UK, AO 13/44/339, viewed at Ancestry.com.

119. Petition of Margaret Draper, Massachusetts, American Loyalist Claims, National Archives, UK, AO 13/44/343, viewed at Ancestry.com.

120. Petition of Margaret Draper, Massachusetts, American Loyalist Claims, National Archives, UK, AO 13/44/351–52, viewed at Ancestry.com.

121. Tremaine, *A Bibliography of Canadian Imprints, 1751–1800*, 666; Gwendolyn Davies, "New Brunswick Loyalist Printers in the Post-War Atlantic World: Cultural Transfer and Cultural Challenges," in *The Loyal Atlantic: Remaking the British Atlantic in the Revolutionary Era*, ed. Jerry Bannister and Liam Riordan (Toronto: University of Toronto Press, 2012), 128–61.

Chapter 6 · Rebuilding Print Networks for the New Nation

1. *Maryland Chronicle, or The Universal Advertiser* (Frederick), January 18, 1786.

2. The translation comes from the Library of Congress catalog record for the *Maryland Chronicle*, https://lccn.loc.gov/sn85025332. On newspaper mottoes, see Jordan E. Taylor, "Eighteenth-Century American Newspaper Mottoes," http://jordanetaylor.com/digital-projects/newspaper-mottoes/.

3. Joseph T. Wheeler, *The Maryland Press, 1777–1790* (Baltimore: Maryland Historical Society, Waverly Press, 1938), 57–64.

4. Alan Taylor, *American Revolutions: A Continental History, 1750–1804* (New York: W. W. Norton, 2016), 313–47.

5. Taylor, *American Revolutions*, 361–66.

6. E. James Ferguson, *The Power of the Purse: A History of American Public Finance, 1776–1790* (Chapel Hill: IEAHC, University of North Carolina Press, 1961), 146–76; Charles Rappleye, *Robert Morris: Financier of the American Revolution* (New York: Simon & Schuster, 2010), 307–57.

7. Terry Bouton, *Taming Democracy: "The People," the Founders, and the Troubled Ending of the American Revolution* (New York: Oxford University Press, 2007); Woody Holton, *Unruly Americans and the Origins of the Constitution* (New York: Hill and Wang, 2007).

8. Four towns lost printing offices between 1782 and 1789: Chatham, New Jersey (printer Shepard Kollock moved to nearby Elizabeth); Westminster, Vermont (the Spooner brothers moved to Windsor); Halifax, North Carolina; and Greenwich, Connecticut.

9. Jessica Choppin Roney, "1776, Viewed from the West," *Journal of the Early Republic* 37, no. 4 (Winter 2017): 655–700; Jack P. Greene, "Colonial History and National History: Reflections on a Continuing Problem," *William and Mary Quarterly*, 3rd ser., 64, no. 2 (April 2007): 235–50; François Furstenberg, "The Significance of the Trans-Appalachian Frontier in Atlantic History," *American Historical Review* 113, no. 3 (June 2008): 647–77.

10. William J. Gilmore, *Reading Becomes a Necessity of Life: Material and Cultural Life in Rural New England, 1780–1835* (Knoxville: University of Tennessee Press, 1989); Jack Larkin, "'Printing is something every village has in it': Rural Printing and Publishing," in *An Extensive Republic: Print, Culture, and Society in the New Nation, 1790–1840*, vol. 2 of *A History of the Book in America*, ed. Robert A. Gross and Mary Kelley (Chapel Hill: University of North Carolina Press, 2010), 145–60.

11. On the early history of Vermont, see Michael Sherman, Gene Sessions, and P. Jeffrey Potash. *Freedom and Unity: A History of Vermont* (Barre: Vermont Historical Society, 2004),

esp. 73–143; Charles T. Morrissey, *Vermont: A Bicentennial History* (New York: W. W. Norton, 1981), 61–105.

12. Todd A. Farmerie, "Episodes in the Life of Anthony Haswell, Postmaster of the Vermont Republic," *Vermont Genealogy* 20 (2015): 167–91; Isaiah Thomas and Boston (MA) Overseers of the Poor, "Document of Indenture: Servant: Haswell, Anthony. Master: Thomas, Isaiah. Town of Master: Boston," Digital Commonwealth, July 23, 1771, http://ark.digitalcommonwealth.org/ark:/50959/1z40m374w.

13. *Vermont Journal, and the Universal Advertiser* (Windsor), August 7, 1783.

14. Esther Littleford Woodworth-Barnes, "Alden Spooner's Autobiography," in *Spooner Saga: Judah Paddock Spooner and His Wife Deborah Douglas of Connecticut and Vermont and Their Descendants; Alden Spooner's Autobiography; Spooner, Douglas, and Jerman Ancestry* (Boston: Newbury Street Press, 1997), 223–64; Gilmore, *Reading Becomes a Necessity of Life*, 195–97; J. Kevin Graffignino, " 'We have long been wishing for a good printer in this vicinity': The State of Vermont, the First East Union and the Dresden Press, 1778–1779," *Vermont History* 47, no. 1 (Winter 1979): 21–36.

15. E. P. Walton, ed., *Records of the Governor and Council of the State of Vermont*, vol. 3 (Montpelier, VT: J. & J.M. Poland, 1875), 45, 392, http://hdl.handle.net/2027/mdp.39015047624005.

16. *Massachusetts Spy*, April 12, 1781.

17. *Massachusetts Spy*, April, 19, 1781. A scrap of the manuscript of the essay exists in the Isaiah Thomas Papers at AAS.

18. Douglas C. McMurtrie, "Pioneer Printing in Georgia," *Georgia Historical Quarterly* 16, no. 2 (1932): 77–113; Rollo G. Silver, "Aprons Instead of Uniforms: The Practice of Printing, 1776–1787," *Proceedings of the American Antiquarian Society* 87, no. 1 (1977): 149–52.

19. On the Society of the Cincinnati, see *Gazette of the State of Georgia*, June 10, 1784, July 15, 1784.

20. On the *Evening Post*, see Clarence S. Brigham, *History and Bibliography of American Newspapers, 1690–1820*, 2 vols. (Worcester, MA: American Antiquarian Society, 1947), 2:931–32.

21. Their claims were similar to the moral authority that many writers claimed in disputes with publishers during the early republic. Michael J. Everton, *The Grand Chorus of Complaint: Authors and the Business Ethics of American Publishing* (New York: Oxford University Press, 2011).

22. *Independent New-York Gazette*, November 22, 1783.

23. *Independent Gazette; or the New-York Journal Revived*, January 10, 1784.

24. The notice appeared in the next number of the *New-York Gazette* on January 31, 1784, the day after his death.

25. *Independent Gazette, or the New-York Journal Revived*, February 19, 1784.

26. *New-York Journal, and State Gazette*, March 18, 1784.

27. Isaiah Thomas, *The History of Printing in America, with a Biography of Printers, and an Account of Newspapers* (Worcester, MA: Isaiah Thomas, Jr., 1810), 1:341–43; Robert E. Burkholder, "Benjamin Edes and John Gill," in *Boston Printers, Publishers and Booksellers: 1640–1800*, ed. Benjamin Franklin V (Boston: G. K. Hall, 1980), 117–34; Maurice R. Cullen, Jr., "Benjamin Edes: Scourge of Tories," *Journalism Quarterly* 51, no. 2 (1974): 213–18; Rollo G. Silver, "Benjamin Edes, Trumpeter of Sedition," *Papers of the Bibliographical Society of America* 47 (1953): 248–68.

28. See Ruma Chopra, *Unnatural Rebellion: Loyalists in New York City during the Revolution*. (Charlottesville: University of Virginia Press, 2011), 188–222.

29. *Independent Gazette*, December 13, 1783.

30. *Independent Gazette*, December 20, 1783.

31. *New-York Gazette, or Journal*, January 17, 1784, referencing a Philadelphia paper of January 3.

32. Ruma Chopra, "Printer Hugh Gaine Crosses and Re-crosses the Hudson," *New York History* 90, no. 4 (2009): 271–85.

33. On Freneau as author, see Paul Leicester Ford, *The Journals of Hugh Gaine, Printer* (New York: Dodd, Mead, 1902), 1:34.

34. *New-Jersey Journal*, January 15, 1783; *Freeman's Journal* (Philadelphia), January 8, 1783; *Massachusetts Spy*, February 6, 1783.

35. *New-Jersey Journal*, February 19, 1783.

36. See Hugh Gaine to Isaiah Thomas, November 10, 1788, Isaiah Thomas Papers, AAS; Hugh Gaine to Isaiah Thomas, January 14, 1789, Hugh Gaine Collection [MS 2958.3677], N-YHS; Gaine to Thomas, January 31, 1789, Typographic Library Collections, Research, Medieval and Renaissance Manuscripts, Rare Book & Manuscript Library, Columbia University in the City of New York.

37. Circular letter addressed to William Young, January 10, 1789, LCP.

38. See, for example, *Royal Gazette* of March 29, 1783; July 19, 1783.

39. *Royal Gazette*, June 28, 1783.

40. *Royal Gazette*, September 24, 1783; October 1, 1783.

41. *Royal Gazette*, November 1, 1783; November 15, 1783; November 19, 1783; November 26, 1783.

42. *Royal Gazette*, October 15, 1783; *Rivington's New-York Gazette*, November 22, 1783.

43. James Rivington to Isaiah Thomas, August 11, 1783, Isaiah Thomas Papers, AAS.

44. "A Whig," *Independent Gazette*, December 27, 1783.

45. *Independent Gazette*, February 5, 1784, listed under news from Worcester, January 15.

46. Isaiah Thomas, will dated May 11, 1784, Isaiah Thomas Papers, AAS.

47. James N. Green, "The Rise of Book Publishing," in Gross and Kelley, *An Extensive Republic*, 75–127.

48. Isaiah Thomas to Hudson and Goodwin, December 8, 1785, Isaiah Thomas Papers, AAS.

49. Green, "The Rise of Book Publishing," 81. See correspondence: Joseph Fry & Sons to Thomas, September 6, 1785; Edmund Fry to Thomas, August 16, 1786; Edmund Fry to Thomas, August 1, 1787, Isaiah Thomas Papers, AAS.

50. John Hancock to Isaiah Thomas, December 2, 1786, Isaiah Thomas Papers, AAS.

51. Ralph Frasca, *Benjamin Franklin's Printing Network: Disseminating Virtue in Early America* (Columbia: University of Missouri Press, 2006), 177–85.

52. Joseph M. Adelman, "Trans-Atlantic Migration and the Printing Trade in Revolutionary America," *Early American Studies, An Interdisciplinary Journal* 11, no. 3 (Fall 2013): 516–44.

53. For more on his life, see James N. Green, *Mathew Carey: Publisher and Patriot* (Philadelphia: Library Company of Philadelphia, 1985), 3–4; Edward C. Carter, II, "The Political Activities of Mathew Carey, Nationalist, 1760–1814" (PhD diss., Bryn Mawr College, 1962); Carter, "Mathew Carey in Ireland, 1760–1784," *Catholic Historical Review* 51, no. 4 (1966): 503–27.

54. Mathew Carey, *Mathew Carey Autobiography* ([Brooklyn]: E. L. Schwaab, 1942), Letter I, 2–3.

55. In 1781, British officials in Ireland forced Carey to flee to France after he published an advertisement for a pamphlet (but not the actual pamphlet) entitled *The Urgent Necessity of an Immediate Repeal of the Whole Penal Code Candidly Considered.* The Library Company of Philadelphia owns the only known extant copy of the pamphlet. See Padhraig Higgins, "Mathew Carey, Catholic Identity, and the Penal Laws," Éire-Ireland 49, nos. 3–4 (Fall–Winter 2014): 176–200.

56. Carter, "The Political Activities of Mathew Carey, Nationalist," 28–37.

57. Carey, *Autobiography*, Letter I, 9; Letter II, 9.

58. Carey, *Autobiography*, Letter II, 11.

59. Mary Pollard, *Dictionary of the Members of the Dublin Book Trade, 1550–1800: Based on the Records of the Guild of St. Luke the Evangelist, Dublin* (London: Bibliographical Society,

2000), 559; Green, *Mathew Carey*, 5–6. The first issue of the *Pennsylvania Evening Herald* listing Spotswood and Talbot as partners was March 26, 1785.

60. *Pennsylvania Evening Herald*, March 26, 1785; April 12, 1785; October 5, 1785.

61. George Washington to Mathew Carey, March 15, 1785, Mathew Carey Papers, AAS.

62. Constitution of North Carolina, December 18, 1776, available at Project Avalon, Yale Law School, http://avalon.law.yale.edu/18th_century/nc07.asp. Other states with free press provisions in their constitutions included Massachusetts, Vermont, Pennsylvania, Maryland, Virginia, South Carolina, and Georgia.

63. Robert W. T. Martin, *The Free and Open Press: The Founding of American Democratic Press Liberty, 1640–1800* (New York: New York University Press, 2001), 100–104.

64. See Joseph M. Adelman and Victoria E.M. Gardner, "News in the Age of Revolution," in *Making News: The Political Economy of Journalism in Britain and America from the Glorious Revolution to the Internet*, ed. Richard R. John and Jonathan Silberstein-Loeb (New York: Oxford University Press, 2015), 47–72.

65. On the tax, see John B. Hench, "Massachusetts Printers and the Commonwealth's Newspaper Advertisement Tax of 1785," *Proceedings of the American Antiquarian Society* 87 (1977): 199–211.

66. John B. Hench, "The Newspaper in a Republic: Boston's *Centinel* and *Chronicle*" (PhD diss., Clark University, 1979), 92–103.

67. "Petition of John Russell printer for relief. Feb 3d, 1786," in John B. Hench, "Massachusetts Printers and the Commonwealth's Newspaper Advertisement Tax of 1785," *Proceedings of the American Antiquarian Society* 87 (1977): 203.

68. Printers also complained often of the scourge of newspaper thieves. See Charles G. Steffen, "Newspapers for Free: The Economies of Newspaper Circulation in the Early Republic," *Journal of the Early Republic* 23, no. 3 (2003), 381–419.

69. *Continental Journal*, May 5, 1785; reprinted in *Independent Ledger*, May 9, 1785; *Essex Journal*, May 11, 1785.

70. *Pennsylvania Evening Herald*, August 3, 1785; *Providence Gazette*, August 13, 1785; *Connecticut Courant*, August 13, 1785; *Carlisle Gazette*, August 17, 1785; *Continental Journal*, August 18, 1785; *Independent Ledger*, August 22, 1785; *American Herald*, August 22, 1785; *Salem Gazette*, August 23, 1785; *Massachusetts Spy*, September 1, 1785; *Falmouth Gazette*, September 3, 1785; *South Carolina Gazette, and Weekly Advertiser*, September 6, 1785; *Vermont Journal*, September 6, 1785.

71. *Pennsylvania Evening Herald*, August 6, 1785; *American Herald*, August 22, 1785; *Salem Gazette*, August 23, 1785; *Continental Journal*, August 25, 1785; *United States Chronicle*, August 25, 1785; *Connecticut Courant*, August 29, 1785; *Hampshire Herald*, August 30, 1785; *Massachusetts Spy*, September 1, 1785; *Falmouth Gazette*, September 3, 1785; *Plymouth Journal*, September 6, 1785.

72. Joanne B. Freeman, "Explaining the Unexplainable: The Cultural Context of the Sedition Act," in *The Democratic Experiment: New Directions in American Political History*, ed. Meg Jacobs, William J. Novak, and Julian E. Zelizer (Princeton, NJ: Princeton University Press, 2003), 20–49.

73. Mathew Carey, *The Plagi-Scurriliad, A Hudibrastic Poem. Dedicated to Colonel Eleazer Oswald*, Early American Imprints, 1st ser., 19540 (Philadelphia: Mathew Carey, 1786).

74. Robert W. T. Martin, *The Free and Open Press: The Founding of American Democratic Press Liberty, 1640–1800* (New York: New York University Press, 2001), 102–4.

75. Saul Cornell, *The Other Founders: Anti-Federalism and the Dissenting Tradition in America, 1788–1828* (Chapel Hill: OIEAHC, University of North Carolina Press, 1999), 128–36.

76. Benjamin H. Irvin, *Clothed in Robes of Sovereignty: The Continental Congress and the People Out of Doors* (New York: Oxford University Press, 2011).

77. Marcus L. Daniel, *Scandal and Civility: Journalism and the Birth of American Democracy* (New York: Oxford University Press, 2009), 155.

78. On Webster, see Joshua Kendall, *The Forgotten Founding Father: Noah Webster's Obsession and the Creation of an American Culture* (New York: Berkley Books, 2010); Jill Lepore, *A Is for American: Letters and Other Characters in the Newly United States* (New York: Alfred A. Knopf, 2002), 15–60; Daniel, *Scandal and Civility*, 148–86.

79. Robb K. Haberman, "Provincial Nationalism: Civic Rivalry in Postrevolutionary American Magazines," *Early American Studies, An Interdisciplinary Journal* 10, no. 1 (Winter 2012): 175–84.

80. George Washington to Mathew Carey, June 25, 1788, in *The Papers of George Washington Digital Edition*, ed. Theodore J. Crackel (Charlottesville: University of Virginia Press, Rotunda, 2008).

81. "Extract of a letter from his Excellency General Washington to the printer of the American Museum" (Philadelphia: Mathew Carey, 1788). The copy sent to Dickinson is owned by LCP [sm #Am 1788 Washi Dickinson 75-13].

82. "Dedication," July 30, 1788, *American Museum* (July 1788), v–vi.

83. On Livington's contributions, see Mathew Carey to William Livingston, September 9, 1788, 1:33–34, Lea & Febiger Records [coll. 227B], HSP; for Pinckney's, see Carey to Pinckney, October 1, 1788, 1:60, Lea & Febiger Records [coll. 227B], HSP. "Subscribers' Names," *American Museum* (July 1789): 5–21.

84. David Ramsay, *The History of the American Revolution*, 2 vols. (Philadelphia: R. Aitken & Son, 1789).

85. Trish Loughran, *The Republic in Print: Print Culture in the Age of U.S. Nation Building, 1770–1870* (New York: Columbia University Press, 2007).

86. Joyce E. Chaplin, ed., *Benjamin Franklin's Autobiography*, Norton Critical Edition (New York: W. W. Norton, 2012), 93.

Conclusion

1. On the newspaper debates about the Constitution, see *DHRC*.

2. Trish Loughran, *The Republic in Print: Print Culture in the Age of U.S. Nation Building, 1770–1870* (New York: Columbia University Press, 2007), 121–31.

3. Saul Cornell, *The Other Founders: Anti-Federalism and the Dissenting Tradition in America, 1788–1828* (Chapel Hill: OIEAHC, University of North Carolina Press, 1999); Pauline Maier, *Ratification: The People Debate the Constitution, 1787–1788* (New York: Simon & Schuster, 2010); Jackson Turner Main, *The Antifederalists: Critics of the Constitution, 1781–1788* (Chapel Hill: Published for the Institute of Early American History and Culture at Williamsburg, VA, by the University of North Carolina Press, 1961).

4. Cornell, *The Other Founders*, 26–34.

5. Letter II, October 9, 1787, in *Observations leading to a fair examination of the system of government proposed by the late Convention; and to several essential and necessary alterations in it. In a number of letters from the Federal Farmer to the Republican* ([New York]: [printed by Thomas Greenleaf], 1787), Early American Imprints, 1st ser., no. 20454, 15.

6. Maier, *Ratification*, 71, 74.

7. Maier, *Ratification*, 82–86; Loughran, *The Republic in Print*, 105–58; Joseph M. Adelman, "Did Hamilton Write Too Much for His Own Good?," *Public Seminar* (blog), November 9, 2017, http://www.publicseminar.org/2017/11/did-hamilton-write-too-much-for-his-own-good/. The essays to be reprinted more than twenty times are Landholder VI and XII, and An American Citizen I, II, and III.

8. Maier, *Ratification*.

9. Maier, *Ratification*; John K. Alexander, *The Selling of the Constitutional Convention: A*

History of News Coverage (Madison, WI: Madison House, 1990); Carol Sue Humphrey, *The Press of the Young Republic, 1783–1833*, History of American Journalism 2 (Westport, CT: Greenwood Press, 1996), 7–11.

10. *New-York Journal*, October 4, 1787, in *DHRC* 13:315.

11. "Argus," *United States Chronicle* (Providence), November 8, 1787, in *DHRC* 13:320–21.

12. *New-York Journal*, October 4, 1787, in *DHRC* 13:315.

13. *Independent Chronicle* (Boston), October 4, 1787, in *DHRC* 13:315.

14. "The Jewel," *Independent Gazetteer* (Philadelphia), November 2, 1787, in *DHRC* 13:320.

15. "A Pennsylvania Mechanic," *Independent Gazetteer* (Philadelphia), October 29, 1787; "Galba," *Independent Gazetteer* (Philadelphia), October 31, 1787, both in *DHRC* 13:319.

16. *Freeman's Journal* (Philadelphia), October 24, 1787, in *DHRC* 13:317.

17. "Detector," *New-York Journal*, October 25, 1787, in *DHRC* 13:318.

18. *Freeman's Journal* (Philadelphia), October, 31, 1787, in *DHRC* 13:320.

19. Jeffrey L. Pasley, "Thomas Greenleaf: Printers and the Struggle for Democratic Politics and Freedom of the Press," in *Revolutionary Founders: Rebels, Radicals, and Reformers in the Making of the Nation*, ed. Alfred F. Young, Gary B. Nash, and Ray Raphael (New York: Alfred A. Knopf, 2011), 355–57.

20. Centinel XI, *Independent Gazetteer* (Philadelphia), January 16, 1788, in *DHRC* 15:389.

21. Cornell, *The Other Founders*.

22. William Goddard to Mathew Carey, February 28, 1788; *Maryland Journal*, February 29, 1788; both in *DHRC* 16:553.

23. "Petition of the Philadelphia Newspaper Printers to the Pennsylvania Assembly," March 20–29, 1788, in *DHRC* 16:563.

24. Main, *Antifederalists*, 250–52.

25. May 7, 1788, *JCC* 34:144.

26. Richard R. John, *Spreading the News: The American Postal System from Franklin to Morse* (Cambridge, MA: Harvard University Press, 1995), 31–42.

27. Maier, *Ratification*, 313.

28. *New-Jersey Journal, and Political Intelligencer* (Elizabethtown), June 11, 1788, reprinting news from Charleston, May 29, 1788.

29. "Founders Online: From James Madison to a Resident of Spotsylvania County, [27 J. . .," January 27, 1789, http://founders.archives.gov/documents/Madison/01-11-02-0313.

30. "Founders Online: Amendments to the Constitution, [8 June] 1789," June 8, 1789, http://founders.archives.gov/documents/Madison/01-12-02-0126.

31. On the determination to add amendments rather than change language, see Mary Sarah Bilder, *Madison's Hand: Revising the Constitutional Convention* (Cambridge, MA: Harvard University Press, 2015), 174–76.

32. Jeffrey L. Pasley, *"A Tyranny of Printers": Newspaper Politics in the Early American Republic* (Charlottesville: University Press of Virginia, 2001); John, *Spreading the News*; Daniel, *Scandal and Civility*; Loughran, *The Republic in Print*; Cornell, *The Other Founders*; David Waldstreicher, *In the Midst of Perpetual Fetes: The Making of American Nationalism, 1776–1820* (Chapel Hill: OIEAHC, University of North Carolina Press, 1997); Rosalind Remer, *Printers and Men of Capital: Philadelphia Book Publishers in the New Republic* (Philadelphia: University of Pennsylvania Press, 1996).

33. Joseph M. Adelman and Victoria E. M. Gardner, "News in the Age of Revolution," in *Making News: The Political Economy of Journalism in Britain and America from the Glorious Revolution to the Internet*, ed. Richard R. John and Jonathan Silberstein-Loeb (New York: Oxford University Press, 2015), 47–72.

34. *Aurora and General Advertiser* (Philadelphia), March 4, 1797; quoted in Pasley, *"Tyranny of Printers,"* 88.

Primary Sources

Research for this volume was supported by a database of 756 printers, editors, and publishers active during the American Revolutionary era, broadly defined as 1756–96. It includes almost exclusively those who were master printers, and therefore excludes those who never advanced past an apprenticeship and those who remained journeymen or laborers throughout their printing careers. I constructed the database using several sources. First among these is the Printers' Card File at the American Antiquarian Society. To supplement those files, I consulted numerous works on bibliography and the history of printing, including Isaiah Thomas, *The History of Printing in America, with a Biography of Printers, and an Account of Newspapers*, 2 vols. (Worcester, MA: Isaiah Thomas, Jr., 1810); Clarence S. Brigham, *History and Bibliography of American Newspapers, 1690–1820*, 2 vols. (Worcester, MA: American Antiquarian Society, 1947); Leona M. Hudak, *Early American Women Printers and Publishers, 1639–1820* (Metuchen, NJ: Scarecrow Press, 1978); Marie Tremaine, *A Bibliography of Canadian Imprints, 1751–1800* (Toronto: University of Toronto Press, 1952); Benjamin Franklin V, ed., *Boston Printers, Publishers, and Booksellers: 1640–1800* (Boston: G. K. Hall, 1980); Frank Cundall, *A History of Printing in Jamaica from 1717 to 1834* (Kingston: Institute of Jamaica, 1935); Howard S. Pactor, *Colonial British Caribbean Newspapers: A Bibliography and Directory* (New York: Greenwood, 1990). I have also consulted numerous monographs and articles on individual printers.

Anyone who works on the history of printing and publishing during the Revolutionary era enjoys the distinct advantage of having access to several online databases of newspapers and other printed materials. The most important of these are America's Historical Newspapers and America's Historical Imprints, produced by Readex. These databases contain thousands of newspapers, pamphlets, almanacs, and broadsides, which made it possible for me to draw comparisons across space and time as well as to track the progress of individual news paragraphs from one town to another. One should not, however, be fooled into thinking that all newspapers are in these databases. Archives continue to make available to researchers an even broader set of newspapers. I relied in particular on the collections at the American Antiquarian Society (which form the basis for much of the Readex digitization project) and the Library Company of Philadelphia not only for newspapers but also for almanacs, pamphlets, broadsides, and the occasional book that had not been digitized. Both archives were also generous in allowing me to view original materials that had been digitized when questions surfaced about the printed document as a material object.

The manuscript records of colonial and Revolutionary era printers are hard to come by. First and foremost, printers used and used up the paper that passed through their offices in manuscript form. Any manuscript source intended for print was handled by multiple people, from the master printer to a journeyman or senior apprentice who might set the type. Nor were these manuscripts particularly intended to be saved—that was the purpose of printing a text. Furthermore, many of the printers in this study left their offices unwillingly during the imperial crisis or the Revolutionary War and left their papers behind, often to be destroyed. Nonetheless, numerous archives contain manuscript records related to the printing trade. The single largest body of correspondence comes from printers related to Benjamin Franklin, who meticulously preserved his writings as well as his associates'. Much of this record is housed at the American Philosophical Society. The American Antiquarian Society, founded by printer and publisher Isaiah Thomas, holds the records of numerous printers from the era, as does the Library Company of Philadelphia (founded by Franklin). Other archives in which I located records include the Massachusetts Historical Society, Houghton Library at Harvard University, the Rhode Island Historical Society, the New-York Historical Society, the New York Public Library, the Historical Society of Pennsylvania, the University of Pennsylvania Libraries, the David Library of the American Revolution, and the National Archives. Records for Loyalist printers are even more difficult to find, but their petitions to the Loyalist Claims Commission are well preserved. Though I was unable to make the trip to Kew to view them in person at the United Kingdom's National Archives, the David Library holds the entire collection on microfilm, and I was able to view records online through Ancestry.com.

The book also relies on the records of various figures important to the Revolution who interacted with printers and publishers or who published frequently. Many of these sources are published through various projects supported by the National Endowment for the Humanities or the National Historical Publications and Records Commission. These include Founders Online (National Archives), https://founders.archives.gov; The *Documentary History of the Ratification of the Constitution Digital Edition*, ed. John P. Kaminski, Gaspare J. Saladino, Richard Leffler, Charles H. Schoenleber, and Margaret A. Hogan (Charlottesville: University of Virginia Press, 2009), http://rotunda.upress.virginia.edu/founders/RNCN; Leonard W. Labaree et al., eds., *The Papers of Benjamin Franklin*, 43 vols. (New Haven: Yale University Press, 1959–present); Paul H. Smith, ed., *Letters of Delegates to Congress, 1774–1789*, 25 vols. (Washington, DC: Library of Congress, 1976–98); Worthington Chauncey Ford, ed., *Journals of the Continental Congress, 1774–1789*, 34 vols. (Washington, DC: U.S. Government Printing Office, 1904–37).

Secondary Sources

Scholars of the colonial and revolutionary periods have been slow to document the networks of printers and the ways in which they mediated the flow of information in mainland North America and throughout the broader British Atlantic world. Much of the work done on colonial printers veers toward and often fully embraces a biography-centered antiquarianism that has long since evaporated in most other historical subfields. Through the 1960s and 1970s, the study of individual printers dominated the field with only scattered acknowledgments that printers indeed maintained connections with one another. The scholarship that does explore relationships among printers has generally focused either on one geographic location or on a single network of printers, in particular the business networks established by the Green family in New England and by Benjamin Franklin throughout the colonies.

Studies of individual printers or their networks include John William Wallace, *An Old Philadelphian, Colonel William Bradford, the Patriot Printer of 1776. Sketches of His Life* (Philadelphia: Sherman & Co., Printers, 1884); Victor H. Paltsits, *John Holt* (New York: New York Public Library, 1920); Leroy Hewlett, "James Rivington, Loyalist Printer, Publisher, Bookseller of the American Revolution, 1724–1802" (PhD diss., University of Michigan, 1958); Ward L. Miner, *William Goddard: Newspaperman* (Durham, NC: Duke University Press, 1962); Alfred Lawrence Lorenz, *Hugh Gaine: Colonial Printer-Editor's Odyssey to Loyalism* (Carbondale: Southern Illinois University Press, 1972); Alan Dyer, *A Biography of James Parker, Colonial Printer* (Troy, NY: Whitston, 1982); Dwight L. Teeter Jr., "John Dunlap: The Political Economy of a Printer's Success," *Journalism Quarterly* 52 (Spring 1975): 3–8, 55. On Franklin as a printer, see J. A. Leo Lemay, *The Life of Benjamin Franklin*, 3 vols. (Philadelphia: University of Pennsylvania Press, 2006–8); James N. Green and Peter Stallybrass, *Benjamin Franklin: Writer and Printer* (New Castle, DE: Oak Knoll Press, 2006); James N. Green, "Benjamin Franklin as Publisher and Bookseller," in *Reappraising Benjamin Franklin: A Bicentennial Perspective*, ed. J. A. Leo Lemay (Newark: University of Delaware Press, 1993); and Ralph Frasca, *Benjamin Franklin's Printing Network: Disseminating Virtue in Early America* (Columbia: University of Missouri

Press, 2006); "Benjamin Franklin's Printing Network," *American Journalism* 5, no. 3 (1988): 145–58; "From Apprentice to Journeyman to Partner: Benjamin Franklin's Workers and the Growth of the Early American Printing Trade," *Pennsylvania Magazine of History and Biography* 114 (1990): 229–48; "'At the Sign of the Bribe Refused': The *Constitutional Courant* and the Stamp Tax, 1765," *New Jersey History* 107 (1989): 21–39. Two exceptions to this trend include Charles Wetherell, who conducted statistical analysis of printers from 1639 to 1783, and Carol Sue Humphrey. See Charles Wetherell, "Brokers of the Word: An Essay in the Social History the Early American Press, 1639–1783" (PhD diss., University of New Hampshire, 1980); Carol Sue Humphrey, *"This Popular Engine": New England Newspapers during the American Revolution, 1775–1789* (Newark: University of Delaware Press, 1992). On the Green family, see Sidney E. Berger, "Innovation and Diversity among the Green Family of Printers" *Printing History* 12, no. 1 (1990): 2–20.

On the importance of print to the American Revolution, see Robert G. Parkinson, *The Common Cause: Creating Race and Nation in the American Revolution* (Chapel Hill: OIEAHC, University of North Carolina Press, 2016); Stephen Botein, "'Meer Mechanics' and an Open Press: The Business and Political Strategies of Colonial American Printers," *Perspectives in American History* 9 (1975): 127–225; Bernard Bailyn, *The Ideological Origins of the American Revolution*, rev. ed. (Cambridge, MA: Belknap Press of Harvard University Press, 1992); T. H. Breen, *The Marketplace of Revolution: How Consumer Politics Shaped American Independence* (New York: Oxford University Press, 2004); Richard D. Brown, *Knowledge Is Power: The Diffusion of Information in Early America, 1700–1865* (New York: Oxford University Press, 1989); Bernard Bailyn and John B. Hench, eds., *The Press and the American Revolution*, (Worcester, MA: American Antiquarian Society, 1980); Philip G. Davidson, *Propaganda and the American Revolution, 1763–1783* (Chapel Hill: University of North Carolina Press, 1941); Humphrey, *"This Popular Engine*; Arthur M. Schlesinger Sr., *Prelude to Independence: The Newspaper War on Britain, 1764–1776* (New York: Alfred A. Knopf, 1957); Eric Slauter, "Reading and Radicalization: Print, Politics, and the American Revolution," *Early American Studies* 8, no. 1 (2010): 5–40.

On newspapers in Revolutionary America, the early United States, and the Atlantic world, see Clarence S. Brigham, *History and Bibliography of American Newspapers, 1690–1820*, 2 vols. (Worcester, MA: American Antiquarian Society, 1947); Charles E. Clark, *The Public Prints: The Newspaper in Anglo-American Culture, 1665–1740* (Oxford: Oxford University Press, 1994); Carol Sue Humphrey, *The American Revolution and the Press: The Promise of Independence* (Evanston, IL: Northwestern University Press, 2013); Uriel Heyd, *Reading Newspapers: Press and Public in Eighteenth-Century Britain and America*, SVEC 2012:3 (Oxford: Voltaire Foundation, 2012); Marcus L. Daniel, *Scandal and Civility: Journalism and the Birth of American Democracy* (New York: Oxford University Press, 2009); William Slauter, "News and Diplomacy in the Age of the American Revolution" (PhD diss., Princeton University, 2007).

On communication more broadly in colonial America and the early United States, see Alejandra Dubcovsky, *Informed Power: Communication in the Early American South* (Cambridge, MA: Harvard University Press, 2016); Katherine Grandjean, *American Passage: The Communications Frontier in Early New England* (Cambridge, MA: Harvard University Press, 2015); Lindsay O'Neill, *The Opened Letter: Networking in the Early Modern British World*, Early Modern Americas (Philadelphia: University of Pennsylvania Press, 2015); Konstantin Dierks, *In My Power: Letter Writing and Communications in Early America* (Philadelphia: University of Pennsylvania Press, 2009); Victoria E. M. Gardner, *The Business of News in England, 1760–1820*, Palgrave Studies in the History of the Media (Houndmills: Palgrave Macmillan, 2016); Ian K. Steele, *The English Atlantic, 1675–1740: An Exploration of Communication and Community* (New York: Oxford University Press, 1986).

On the history of the book, see Hugh Amory and David D. Hall, eds., *The Colonial Book in the Atlantic World*, vol. 1 of *A History of the Book in America* (Cambridge: Cambridge University

Press, 2000); Robert A. Gross and Mary Kelley, eds., *An Extensive Republic: Print, Culture, and Society in the New Nation, 1790–1840*, vol. 2 of *A History of the Book in America* (Chapel Hill: University of North Carolina Press, 2010); Robert Darnton, "What Is the History of Books?," *Daedalus* 111, no. 3 (1982): 65–83; Larzer Ziff, *Writing in the New Nation: Prose, Print, and Politics in the Early United States* (New Haven, CT: Yale University Press, 1991); David S. Shields, *Civil Tongues & Polite Letters in British America* (Chapel Hill: OIEAHC, University of North Carolina Press, 1997); Joan Shelley Rubin, "What Is the History of the History of Books?," *Journal of American History* 90 (2003): 555–75; David D. Hall, *Worlds of Wonder, Days of Judgment: Popular Religious Belief in Early New England* (Cambridge, MA: Harvard University Press, 1990).

On the concept of the public sphere, see Jürgen Habermas, *The Structural Transformation of the Public Sphere: An Inquiry into a Category of Bourgeois Society*, trans. Thomas Burger (Cambridge, MA: MIT Press, 1989); Christopher Grasso, *A Speaking Aristocracy: Transforming Public Discourse in Eighteenth-Century Connecticut* (Chapel Hill: OIEAHC, University of North Carolina Press, 1999); Michael Warner, *The Letters of the Republic: Publication and the Public Sphere in Eighteenth-Century America* (Cambridge, MA: Harvard University Press, 1990); Craig J. Calhoun, ed., *Habermas and the Public Sphere* (Cambridge, MA: MIT Press, 1992). On the concept of the nation as an "imagined community," see Benedict Anderson, *Imagined Communities: Reflections on the Origin and Spread of Nationalism*, rev. ed. (London: New York: Verso, 2006); Trish Loughran, *The Republic in Print: Print Culture in the Age of U.S. Nation Building, 1770–1870* (New York: Columbia University Press, 2007).

Scholarship on political culture in Revolutionary America and the early United States has had an enormous impact on the field in the past twenty years. The scholars I have met and interacted with through the Society for Historians of the Early American Republic (SHEAR) have influenced this book in many ways. See, for example, Pauline Maier, *Ratification: The People Debate the Constitution, 1787–1788* (New York: Simon & Schuster, 2010); Jeffrey L. Pasley, *"The Tyranny of Printers": Newspaper Politics in the Early American Republic*, Jeffersonian America (Charlottesville: University of Virginia Press, 2001); Joanne B. Freeman, *Affairs of Honor: National Politics in the New Republic* (New Haven, CT: Yale University Press, 2001); Richard R. John, *Spreading the News: The American Postal System from Franklin to Morse* (Cambridge, MA: Harvard University Press, 1995); Saul Cornell, *The Other Founders: Anti-Federalism and the Dissenting Tradition in America, 1788–1828* (Chapel Hill: OIEAHC, University of North Carolina Press, 1999); David Waldstreicher, *In the Midst of Perpetual Fetes: The Making of American Nationalism, 1776–1820* (Chapel Hill: OIEAHC, University of North Carolina Press, 1997); Pauline Maier, *American Scripture: Making the Declaration of Independence* (New York: Alfred A. Knopf, 1997).

INDEX

Page numbers in *italics* refer to illustrations. The letter "t" following a page number denotes a table.